Children and Nature

Children and Nature

Psychological, Sociocultural, and Evolutionary Investigations

Edited by Peter H. Kahn, Jr. and Stephen R. Kellert

The MIT Press
Cambridge, Massachusetts
London, England

This book was set in Sabon by SNP Best-set Typesetter Ltd., Hong Kong. Printed on recycled paper and bound in the United States of America.

Library of Congress Cataloging-in-Publication Data

Children and nature : psychological, sociocultural, and evolutionary investigations / edited by Peter H. Kahn, Jr. and Stephen R. Kellert.
p. cm.
Includes bibliographical references and index.
ISBN 978-0-262-11267-3 (hc : alk. paper)—ISBN 978-0-262-61175-6 (pbk. : alk. paper)
1. Nature—Psychological aspects. 2. Environment and children. I. Kahn, Peter H. II. Kellert, Stephen R.

BF353.5.N37 C47 2002
305.23—dc21 2001057912

10 9 8 7 6

Contents

Introduction vii
Peter H. Kahn, Jr. and Stephen R. Kellert

1 **The Primate Relationship with Nature: Biophilia as a General Pattern 1**
Peter Verbeek and Frans B. M. de Waal

2 **The Ecological World of Children 29**
Judith H. Heerwagen and Gordon H. Orians

3 **The Development of Folkbiology: A Cognitive Science Perspective on Children's Understanding of the Biological World 65**
John D. Coley, Gregg E. A. Solomon, and Patrick Shafto

4 **Children's Affiliations with Nature: Structure, Development, and the Problem of Environmental Generational Amnesia 93**
Peter H. Kahn, Jr.

5 **Experiencing Nature: Affective, Cognitive, and Evaluative Development in Children 117**
Stephen R. Kellert

6 **Animals as Links toward Developing Caring Relationships with the Natural World 153**
Olin Eugene Myers, Jr. and Carol D. Saunders

7 **Animals in Therapeutic Education: Guides into the Liminal State 179**
Aaron Katcher

8 Spots of Time: Manifold Ways of Being in Nature in Childhood 199
Louise Chawla

9 Adolescents and the Natural Environment: A Time Out? 227
Rachel Kaplan and Stephen Kaplan

10 Adolescents and Ecological Identity: Attending to Wild Nature 259
Cynthia Thomashow

11 Political Economy and the Ecology of Childhood 279
David W. Orr

12 Eden in a Vacant Lot: Special Places, Species, and Kids in the Neighborhood of Life 305
Robert Michael Pyle

List of Contributors 329
Name Index 331
Subject Index 341

Introduction

Peter H. Kahn, Jr. and Stephen R. Kellert

For much of human evolution, the natural world constituted one of the most important contexts children encountered during their critical years of maturation. It would not be too bold to assert that direct and indirect experience of nature has been and may possibly remain a critical component in human physical, emotional, intellectual, and even moral development. Despite this possibility, our scientific knowledge of the impact and significance of nature during varying stages of childhood is remarkably sparse. For example, we remain largely uninformed about the following questions:

• Do young children form deep connections with the natural world, or is that idea actually a myth?

• What are the evolutionary origins of children's relationships with nature?

• How do children form environmental commitments and sensibilities and reason about environmental issues?

• Do animals provide a means by which children come to care about nonsentient nature? Or about other humans?

• Do developmental needs and thresholds exist relative to the quantity and quality of children's contact with the natural world?

• What is the relative significance of direct, indirect, and symbolic experiences of nature during childhood?

• Does it matter that many children today encounter substantially fewer opportunities for direct experience with healthy natural systems?

• What is the significance of increasing children's exposure to nature through technologically mediated interactions (as occur with television and computers)?

This book offers a partial response to these questions and others regarding children's relationship to and dependency on the natural world.

More than two decades ago, scholars such as Edith Cobb (1977/1993), Rachel Carson (1962/1994), and Harold Searles (1959), writing from different disciplinary traditions, argued that a child's experience of nature exerted a crucial and irreplaceable effect on physical, cognitive, and emotional development. Despite these assertions, we continue to rely largely on anecdote and assumption rather than on systematic examination and well-articulated theoretical formulation for our understanding of the role of nature in child development. Even partial answers to the questions posed above could have enormous significance in areas such as child rearing, education, land-use planning, and the design of the natural and human-built environment.

Toward this end, we have chosen contributors from diverse fields, including cognitive science, developmental psychology, ecology, education, environmental studies, evolutionary psychology, political science, primatology, psychiatry, and social psychology. In turn, we have organized this volume around three broad perspectives. The first perspective emphasizes the evolutionary significance of nature during childhood. Although this perspective is the least developed of the three in this volume (and thus placed last in the subtitle), we take as a starting point that humans are biological beings with an evolutionary history and that any comprehensive account of children and nature must dovetail with and then build from this evolutionary basis. The second perspective is psychological. This perspective emphasizes how children form conceptions, values, and sympathies toward the natural world and how contact with nature affects children's physical and mental development. The third perspective is sociocultural. This perspective emphasizes educational and political consequences arising from the changing quality and quantity of childhood experiences of the natural world in modern society.

The overall volume proceeds as follows. Peter Verbeek and Frans B. M. de Waal (chapter 1) focus on the nonhuman primate relationship with nature as a way to lay the evolutionary foundations for this volume.

In their chapter, they draw on the idea of biophilia—the proposition that humans have a fundamental, genetically based human need and propensity to affiliate with nature. Verbeek and de Waal suggest that if the idea of biophilia has merit, then it should be possible to detect aspects of biophilia in the natural behavior of nonhuman primates such as monkeys and apes, which have walked the evolutionary road with humans for a long time and have faced similar problems in finding their way in nature, literally and metaphorically.

In supporting their thesis, Verbeek and de Waal draw on an impressive body of empirical research with nonhuman primates. And they show us some remarkable findings. They show us, for example, that nonhuman primates are skilled at exploiting their natural habitat. Chimpanzees, for example, use tools to manipulate nature (such as "hammers" and "anvils" to crack open palm nuts), and great apes learn to use certain plants for medicinal purposes. Moreover—as the biophilia hypothesis predicts—the emotions of nonhuman primates appear deeply woven into their explorations of the natural world. For example, a foraging party of chimpanzees that discovers an abundant food source may respond by hooting and drumming on trees, attracting other members of the larger community to a scene that some researchers have described as a "carnival." Or when female chimpanzees are traveling together, a subordinate individual may touch or embrace a dominant travel companion before rushing toward the desired food. Nonhuman primates also exhibit what may be termed a sense of wonder in the natural world. For example, Verbeek once observed a group of young mandrills cluster tightly around an adult male to inspect a toad that was not doing a very good job at playing dead. The infants and juveniles in the group appeared transfixed by the trespassing amphibian (which, after close inspection, the mandrills let go, apparently unharmed). Verbeek and de Waal comment: "If an early sense of wonder predicts the good naturalist, these mandrill youngsters seemed fit for a career in field biology."

Judith H. Heerwagen and Gordon H. Orians (chapter 2) extend the evolutionary account directly into the lives of children. Specifically, they characterize the ecological relationships of children and predict developmental patterns of behavior based on an evolutionary perspective.

Then they test their predictions with published data. For example, Heerwagen and Orians suggest that in our evolutionary history infants who preferred objects close at hand to objects at a distance would stay closer to their care givers and be better protected, thus conferring a genetic advantage. Indeed, the child developmental literature supports the corresponding prediction. For while older infants and young children can see distant objects relatively well, they do not attend to them. Another example delves into a widely studied developmental phenomenon in attachment theory—that infants' fear of strangers typically begins to develop at around seven months, peaks at about one year, and continues until 18 to 24 months. Heerwagen and Orians predict that an infant's fear of strangers should be more intense for strange males than for strange females because of the greater potential for harm associated with aggression by unknown and unrelated males. Again, the child developmental literature supports this prediction. What is especially elegant here is that Heerwagen and Orians provide a completely reasonable explanation for a seemingly unintuitive finding—moreover, for a finding that has remained poorly explained in the developmental psychological literature. Finally, Heerwagen and Orians raise concerns about the effects of video technology on children's relationship with nature.

Heerwagen and Orians thus offer an insightful method toward validating evolutionary hypotheses. At the same time, it is sometimes difficult using their method to distinguish predictions from post hoc accounts of our evolutionary heritage. For example, Heerwagen and Orians note that when an infant begins to walk, he or she increases the likelihood of straying outside the caretaker's field of view. Accordingly, Heerwagen and Orians predict (and find) that at this point in development mouthing of novel objects rapidly diminishes (thus decreasing the infant's chance of ingesting harmful substances). But one could just as easily predict that evolution would bestow on the infant who begins to walk adequate enough cognition to guide his or her own choice of food items. Or one could just as easily predict that evolution would bestow on an infant a heartier biological digestive system that could make light duty of otherwise toxic foods. How does one know which "prediction" to make except by a post hoc analysis? That said, both Verbeek and de Waals (chapter 1) and Heerwagen and Orians (chapter 2) contribute to the broader evolutionary biological program that has begun—with a good deal of success—to lay theoretical foundations across diverse disciplines (Barkow, Cosmides & Tooby, 1992; Wilson, 1998).

John D. Coley, Gregg E. A. Solomon, and Patrick Shafto (chapter 3) provide a further bridge between the biological and psychological perspectives. They show us what cognitive science has to say about the development of "folkbiology"—how children understand, classify, reason about, and explain the world of plants and animals. As Coley et al. review the literature, they also pursue a difficult and foundational theoretical question in child development—namely, what exactly develops? Coley et al. offer two options. One is that the organization of folkbiological knowledge undergoes quantitative change, whereby development is understood as the accumulation of knowledge. Another is that folkbiological knowledge undergoes qualitative change, by which Coley et al. mean that one worldview is replaced by another (presumably more veridical) worldview. In their reading of the evidence, both options have merit, and accordingly they offer a mixed model of development. As a case in point, while children share many elements of adults' understandings of inheritance, illness, and growth, over time these understandings become more interrelated and causally coherent.

Instead of viewing conceptual change as involving the overthrow of one worldview with another, Peter H. Kahn, Jr. (chapter 4) draws on structural-developmental theory that emphasizes the idea of hierarchical integration: that earlier forms of knowledge are not lost in development but are embedded and reworked—transformed—into more comprehensive ways of understanding the world and acting upon it. Based on this theoretical approach, Kahn and his colleagues interviewed children in diverse locations, ranging from an inner-city African American community in Houston, Texas, to the Brazilian Amazon, to Lisbon, Portugal, about their environmental moral reasoning and values. Kahn and his colleagues found evidence for the universality of two overarching environmental moral orientations—anthropocentric and biocentric. Anthropocentric reasoning is based on how effects to the environment affect human beings, including appeals to human welfare, personal interests, and aesthetics. Biocentric reasoning is based on how the natural environment has moral standing that is at least partly independent of its value as a human commodity, including appeals that nature has rights or has intrinsic value. Kahn suggests that through development biocentric reasoning may hierarchically integrate anthropocentric reasoning.

Kahn also articulates what he views as one of the most pressing and unrecognized problems of our age—the problem of environmental generational amnesia. The idea here is that people take the natural environment they encounter during childhood as the norm against which they measure environmental degradation later in their life. With each ensuing generation, the amount of environmental degradation increases, but each generation takes that degraded condition as the nondegraded condition—as the normal experience. The upside of environmental generational amnesia is that each generation starts afresh, unencumbered mentally by the environmental misdeeds of previous generations. But the downside is enormous, for each of us has difficulty understanding in a direct, experiential way that nature as experienced in our childhood is not the norm but is already environmentally degraded. Thus, according to Kahn, environmental generational amnesia helps provide a psychological account of how our world has moved toward its environmentally precarious state.

Stephen R. Kellert (chapter 5) examines the effect of contact with nature on physical and mental development, especially during middle childhood and early adolescence. In approaching this subject, Kellert emphasizes the need to distinguish between direct, indirect, and vicarious experience of nature in examining impacts on child development. Direct experience involves actual physical contact with creatures and habitats largely independent of human input and control. Indirect experience also includes actual physical contact but in a largely restricted, regulated, and constructed human context. Vicarious experience of the natural world involves realistic as well as symbolic and fantastic representations of nature. Kellert further distinguishes between affective, cognitive, and values-related development, invokes the concept of biophilia, and then provides a related typology of inherent tendencies to value the natural world to explain the significance of varying childhood experiences of nature at different ages or stages in personality and character formation.

Finally, Kellert discusses the likely developmental impacts on contemporary children of the apparent decline and impoverishment of direct contact with healthy natural process and diversity and the seemingly concurrent increase in indirect and especially vicarious experiences of the natural world. While reviewing data suggesting positive aspects of increasing organized programs and mass communications exposure to the natural world, Kellert suggests that these experiences do not adequately compensate for diminishing direct encounters with nearby and familiar natural environments. Kellert concludes that the child's direct and ongoing experience of accessible nature is an essential, critical, and irreplaceable dimension of healthy maturation and development.

The concept of nature encompasses plants, objects (such as rocks), events (such as storms), and of course animals. Indeed, some of the literature suggests that the formative importance of animals may be particularly pronounced during early and middle childhood (Beck & Katcher, 1996; Kahn, 1999; Kellert, 1996, 1997; Levi-Strauss, 1970; Searles, 1959; Shepard, 1978, 1996). Many reasons may account for this effect, including the familiarity of certain animals to our own species, assumptions regarding sentience, the capacity for movement, analogous bodily features, and phylogenetic and morphological similarities between people and vertebrate animals.

Olin Eugene Myers, Jr. and Carol D. Saunders (chapter 6) build on this literature and show how caring for animals provides a (perhaps essential) link by which children develop caring relationships with the natural world. Their argument can be viewed as comprising two parts. First, Myers and Saunders provide a developmental account of how children come to care for animals. Namely, very young children begin to understand that animals display four properties that remain constant across many different interactions: *agency* (a dog decides to eat and acts accordingly), *affectivity* (a dog appears to enjoy playing with the child), *coherence* (a dog is able to coordinate its movements in response to the child's actions), and *continuity* (the dog's repeated interactions become regularized into a relationship with the child). Such understandings make it possible for children to recognize that animals have their own subjective states and can have correlative interests in interacting with the child ("my dog wants to play with me"). These cognitive underpinnings, in turn, make possible the development of caring for animals, which Myers and Saunders view as a natural outgrowth of intimate relationships with individual animals. Such natural caring, however, can fall short. After all, what about animals that children do not know personally—a dog across town or macaques in Indonesia? Presumably such animals also deserve moral consideration. So now comes the second part of their argument. And the key idea here—the link between caring for animals and the natural world—in effect builds from structural-developmental theory (chapter 4). For Myers and Saunders suggest that through interaction with people and animals (and increased environmental knowledge), children come to recognize limitations of a moral perspective based on natural caring (that it lacks impartiality, for example) and thus construct more generalized concepts of care for animals in general and the natural world as a whole.

Myers and Saunders's research populations comprised "typical" children—the type, for example, who attend a suburban public school or visit a large metropolitan zoo. But might animals have a special role to play in therapeutic situations? Aaron Katcher (chapter 7) brings his expertise as a child psychiatrist to bear on this question. In a variety of residential treatment programs, Katcher has worked with children diagnosed with autism, developmental disorder, attention-deficit hyperactivity disorder, conduct disorder, and oppositional-defiant disorder. Katcher found that such children persisted in learning the skills and information necessary for them to handle the animals. Through interaction with animals, these children also demonstrated an increase in attention span, a decrease in hostile and aggressive behavior, and an increase in cooperative behavior. Indeed, the skill and care these children displayed in handling and caring for the animals led visitors frequently to ask, "Why are these children in residential treatment?" Moreover, many of the benefits that accrued to these children have been found to generalize to children and adults at large. Thus Katcher asks the difficult question: How does contact with animals change people's behavior? Here Katcher draws on theories initiated by anthropologist Victor Turner and psychoanalyst D. W. Winnecott to suggest that because animals show intentional behavior but cannot contradict the virtuous attributes that we find lacking in fellow human beings, animals create a "liminal" state—one that intensifies good feelings, bonding, and a sense of community.

Louise Chawla (chapter 8) suggests that the scientific approach to the study of children and nature all too often emphasizes cognition at the expense of our deepest levels of connection with the natural world. In response, she brings to the foreground and then integrates two perspectives—the romantic legacy, as typified in the poems of William Wordsworth, and the psychological theory of Jean Gebser. From Wordsworth, Chawla draws on the idea of "spots of time"—special moments that children have with nature that embody tenderness, love, responsive caretaking, and what Chawla calls "patterns of divinity." Then from Gebser, Chawla argues that real wisdom (what Gebser calls "integral wisdom") requires that ways of being in the past (such as ways that involved "patterns of divinity") remain accessible to the present. Thus, Chawla offers a counterpoint to the structural-developmental idea of hierarchical integration (chapter 4). For in her view, childhood ways of knowing nature are not inadequate compared to adult ways but are different, and adult wisdom emerges by accessing them. Or in the words she quotes from Wordsworth: "There are in our existence spots of time, that with distinct pre-eminence retain / A renovating virtue."

Rachel Kaplan and Stephen Kaplan (chapter 9) move the discussion from childhood to adolescence. And they take up a puzzling finding in the research literature that seems to suggest that adolescents, compared to younger and older groups, prefer nature less and developed areas (like shopping malls) more. As they analyze the literature, they find qualified support for this proposition. On the one hand, there does appear to be a "time out" in the adolescent relationship with nature. This time out is driven partly by an intensity that adolescents bring to their desire for activities that convey excitement, for interaction with their peers, for peer acceptance, and for establishing autonomy. The Kaplans then provide a sophisticated discussion of how such desires have both cultural and evolutionary foundations. But the Kaplans also check the initial proposition, for in their careful reading of the literature it becomes apparent that adolescents very much appreciate natural places, both within their community and often in the wild. Thus the Kaplans ask the question: How can educators involve adolescents with nature in ways that also effectively support the need adolescents have for social interaction?

It is this question that Cynthia Thomashow (chapter 10) answers. Like the Kaplans, Thomashow recognizes the uniqueness of the adolescent period in development. And, as an educator, she is all too aware of how the adolescent stance—with its focus on autonomy and peer acceptance—can lead to heated conflicts with teachers (and parents) and often rejection of the educational enterprise. In response, based on the construct of ecological identity, Thomashow provides techniques for bringing ecological thought and an affiliation with nature into the learning process. In particular, Thomashow describes three school-based programs that have successfully integrated ecological thinking into the educational experience of teens. The first engaged high school students in the management of public lands. The second engaged high school students in the protection of a wildlife sanctuary. The third engaged junior high school students in the design of an exhibit at a metropolitan zoo. Through her descriptions, we traverse the edges of the adolescent world and discover how to attend to what Thomashow calls the essential wild nature of adolescent development.

As the Kaplans suggest, fundamental changes have occurred in children's relationships to nature in modern, and especially urban, society. Moreover, expanding human populations, accelerating land and resource consumption, growing urbanization, and ecological degradation all have resulted in significant declines in opportunities for children to encounter healthy and abundant natural systems. Against this backdrop, David W. Orr (chapter 11) engages in a meditation on larger political and ecological patterns and their effects on children. Modern political economies, according to Orr, have led to promiscuous industrial pollution, junk diets that corporations foist on children though insidious advertising, capitalistic consumption that works best when children stay indoors in malls and in front of televisions or computer screens, the subjugation of children to hundreds of harmful chemicals that threaten children's future ability to procreate, the conditions by which children on average can recognize over 1,000 corporate logos but only a handful of plants and animals native to their places, biotic impoverishment, climatic change, the undermining of millions of years of evolution, and the demise of children's rightful heritage to live intimately with the natural world.

Although Orr draws on research findings to argue his case, his characterizations of the problems should be read—in our view—as provocative hypotheses, not established facts. Of course, we are aware that Orr himself would find this very caveat maddeningly conservative. He in effect asks in his chapter why we quibble about this fact or that fact when the overarching global problems are all too obvious.

Robert Michael Pyle (chapter 12) concludes this volume with an eloquent personal and professional exploration of the importance during childhood of direct experience with ordinary and nearby natural areas. Such areas, according to Pyle, lead children along the naturalist's path and toward increasing intimacy with nature. But according to Pyle, "For special places to work their magic on kids, they need to be able to do some clamber and damage. They need to be free to climb trees, muck about, catch things, and get wet—above all, to leave the trail." Thus Pyle emphasizes the "vacant lot"—open ground, a creek, some scrap of the wild. Such areas protect us from what Pyle calls the extinction of experience whereby lack of interaction with rich ecosystems leads to lack of concern for their protection, which leads to further lack of interactions. The extinction of experience is thus a cycle whereby impoverishment begets greater impoverishment. Pyle further considers the potentially compensating effects of increasing formal educational and media-based contacts with nature but remains unconvinced: "Just as real life does not consist primarily of car chases and exploding buildings, quotidian nature is much more about grasshoppers in the pigweed than it is rhinos mating on a pixilated screen." Once we admit the primacy of untended ground in our physical and cultural landscape, then Pyle says that we must take action, and he offers practical suggestions for getting started.

Our primary objective in this volume is to provide scientific investigations into the study of children and nature. Yet at this point, we should say a few words about how we understand the scientific enterprise. To do so, we draw on Machado, Lourenço, and Silva (2000), who distinguish between three kinds of scientific investigations—theoretical, conceptual, and empirical. Theoretical investigations aim at developing a set of principles that can explain empirical regularities or phenomena

(Einstein's theory of relativity or the theory of natural selection). Wilson's (1984) account of "biophilia"—the proposition that humans have a genetic predisposition to affiliate with life and lifelike processes—is in this sense aimed at theory building. In turn, conceptual investigations have at their core the concepts of the theory—their meanings, coherency, logic, and intelligibility. For example, when critics have charged that the construct of biophilia remains too broad to be meaningful or when that charge is in turn answered, the debate is focused on conceptual investigations. Finally, empirical investigations refer to the gathering of data. We mention empirical investigations last, not because they are the least important but because in our view scientists—especially social scientists—often move too quickly to collect data without adequate theoretical and conceptual justification. Equally problematic, the investigations often stop there. In short, we view this volume as contributing to the scientific understanding of children and nature. But we aim for science robustly conceived to embrace the dialectic between theory, concepts, and empirical data.

We recognize that no single book can adequately cover a topic as difficult, multidisciplinary, and relatively unstudied as this one. Nonetheless, the chapters herein provide a start. If our volume has been successful, it will stimulate others, particularly young scholars, to delve deeply into this immensely important topic, one that strikes at the core of what it means to be fully and functionally human. We also hope that this volume will help generate the understanding and concern necessary to motivate societal change. As the chapters in this volume show, there exists a critical and irreplaceable role of nature for all children, whether they reside in urban or nonurban, industrially developed or less developed, Western or non-Western areas. This central finding should cause educators and decision makers to support endeavors that seek to improve and increase opportunities for children to experience nature in intimate, ongoing, and satisfying ways as a core objective of modern society.

References

Barkow, J. H., Cosmides, L., & Tooby, J. (Eds.). (1992). *The adapted mind: Evolutionary psychology and the generation of culture.* New York: Oxford University Press.

Beck, A., & Katcher, A. (1996). *Between pets and people.* West Lafayette, IN: Purdue University Press.

Carson, R. (1994). *Silent spring.* Boston: Houghton Mifflin. (Original work published 1962).

Cobb, E. (1993). *The ecology of imagination in childhood.* Dallas, TX: Spring. (Original work published 1977).

Kahn, P. H., Jr. (1999). *The human relationship with nature: Development and culture.* Cambridge, MA: MIT Press.

Kellert, S. R. (1996). *The value of life.* Washington, DC: Island Press.

Kellert, S. R. (1997). *Kinship to mastery: Biophilia in human evolution and development.* Washington, DC: Island Press.

Levi-Strauss, C. (1970). *The savage mind.* Chicago: University of Chicago Press.

Machado, A., Lourenço, O., & Silva, F. J. (2000). Facts, concepts, and theories: The shape of psychology's epistemic triangle. *Behavior and Philosophy, 28,* 1–40.

Searles, H. F. (1959). *The nonhuman environment.* New York: International Universities Press.

Shepard, P. (1978). *Thinking animals.* New York: Viking.

Shepard, P. (1996). *The others: How animals made us human.* Washington, DC: Island Press.

Wilson, E. O. (1984). *Biophilia.* Cambridge, MA: Harvard University Press.

Wilson, E. O. (1998). *Consilience.* New York: Knopf.

Children and Nature

1

The Primate Relationship with Nature: Biophilia as a General Pattern

Peter Verbeek and Frans B. M. de Waal

How can we best describe our evolving place in nature? To what extent does our relationship with nature affect us, both as a species and as individuals? Both questions have been approached from the perspective of what E. O. Wilson and others have called *biophilia* (Kellert & Wilson, 1993; Wilson, 1984, 1993). Wilson (1984) initially proposed biophilia as an innate tendency to affiliate with natural things. He later specified the role of emotion and suggested that when we encounter living things, we experience emotions ranging from "attraction to aversion, from awe to indifference, [and] from peacefulness to fear-driven anxiety" (Wilson, 1993, p. 31). These emotions are thought to be linked to adaptive learning rules that govern how we learn about and from nature. During evolutionary history this complex of emotions and learning rules has helped us survive and become who we are. One way of interpreting biophilia is to consider it a functional subunit of our "adapted mind," not unlike, say, our natural ability for language or culture (cf. Barkow, Cosmides & Tooby, 1992).

We have two goals for this chapter. First, we want to review aspects of the primate relationship with nature from the perspective of the construct of biophilia. If biophilia is rooted in our evolutionary past, it is reasonable to assume that it may have originated in an ancestry we share with currently living primates. If so, we should be able to detect aspects of it in the natural behavior of extant monkeys and apes. Second, we want to show how studying the nonhuman primate relationship with nature can tell us something about our own relationship with nature, in particular how it manifests itself in childhood. Haraway (1991) has suggested that primatology, the science of nonhuman primates, may either be a source of insight into our own behavior or a source of illusion. She proposes that the issue rests on the type of mirror through which we chose to view ourselves. In this chapter we look at biophilia in our closest living relatives in the hope that it may bring into focus some of the intricacies of children's developing relationship with nature.

Throughout this chapter we visit with primatologists and other naturalists, zoo visitors, children of various ages, and, most important, a variety of free-ranging and captive primates. We start with reflections on how biophilia may manifest itself when we meet our primate kin, either in our capacity as professional primatologists or, for example, during a family outing to the zoo. We then highlight some of the literature on children's relationships with nature to link it to our approach. Finally, we discuss aspects of the primate relationship with nature, drawing on examples that not only blend emotion and learning but also reflect a degree of mastery and kinship that we believe may be characteristic of primate biophilia.

Biophilia

Primatologists seem an empathic lot. For one, they are inclined to embellish their observations with emotional reflections. Take the renowned chimpanzee expert Jane Goodall. She confided that her heart missed several beats when she first came face to face with the now famous Gombe chimpanzees (van Lawick-Goodall, 1971). One of us suggested elsewhere that when we make eye contact with an anthropoid ape, "we can feel a powerful personality that resembles our own, both emotionally and mentally" (de Waal & Lanting, 1997, p. 1). Primatologists are not alone in blending a sense of wonder with systematic inquiry and empathy with reason. Jonas Salk has spoken of becoming a virus and imagining the response that the immune system would make. And behavioral biology's grand master of observation, Konrad Lorenz, emphasized the importance of appreciating an animal's beauty for truly understanding its behavior. Finding joy in just looking at animals allows us to have the patience to observe them long enough to see something of interest to science.

Evidence that one doesn't have to be a professional primatologist to be moved when face to face with primate kin comes from a recent study conducted at Chicago's Brookfield Zoo. Researchers asked visitors about their reactions to the zoo's monkeys and apes, and many commented on being touched by the humanlike expressions and behaviors of the primates (Saunders & Wood, 1992), especially their mother-infant interactions and their use of their hands to manipulate food. Facial expressions struck a cord too. Respondents were reluctant to identify aspects of primates that they did not like; some commented on sexual behavior or alluded to aggression, but most avoided these areas. When the researchers asked for comments on the statement "humans are primates," the common reaction was one of surprise; only a few of the respondents accepted the statement as a matter of fact. Many attempted to distance themselves from primates—for instance, by citing differences in intellect.

In addition to suggesting the need for more effective education about our evolutionary place among primates, Saunders and Woods's study shows that meeting our primate kin is likely to stir emotions in us. We may alternate between attraction and repulsion, but meeting our primate cousins most likely will not leave us cold. These emotions are most certainly a manifestation of biophilia. They appear widespread among us and, as Wilson (1993) predicts, range from attraction to aversion. Biophilic reactions in response to our primate kin may be especially potent since both in appearance and behavior monkeys and apes remind us of ourselves. Considering the extent to which we tend to be enthralled with ourselves, our emotional fascination with our primate kin thus comes as no great surprise.

Biophilia is more than a fascination with primates, however. An increasing body of research suggests that nature's psychological pull on us spans a range of contexts and incorporates cognitive as well as emotional correlates. Kellert (1996, 1997), for instance, delineated a set of universal values that reflect both our kinship with nature and our relative mastery of it. Kahn (1997, 1999) showed how some of these values may take hold during childhood and adolescence. Other studies have provided evidence for the developmental roots of biophilia. For instance, several studies indicate that young children may be especially attuned[1] to natural kinds. Gelman (1990) and Gelman and Markman (1987), for instance, showed that being a member of a natural kind carried more inferential weight for young children than being perceptually similar. Wohlwill (1983) demonstrated that children as young as six years of age spontaneously sort natural from human-made stimuli. Finally, Atran (1990) showed, in a related finding, that principles that adults use to spontaneously categorize plants and animals appear to be the same cross-culturally.

Lovelock (1991) proposes that our recognition of living things, both animal and vegetable, is instant and automatic and that our fellow creatures in the animal world appear to have the same facility. He adds that this powerful but unconscious recognition no doubt evolved as a survival factor. As Lovelock (1991, p. 7) puts it: "Anything living may be edible, lethal, friendly, aggressive, or a potential mate—all questions of prime significance for our welfare and continued existence." Thus Lovelock reminds us that the ability to recognize living things has comprehensive adaptive value. We would like to add that being able to recognize living things is a necessary, but not sufficient, step toward being attracted to them. As such, the ability to recognize living things may be an ancient (in evolutionary terms) component part of biophilia.[2] Nonhuman primates are not special in demonstrating the ability to recognize living things. However, within the constraints of their particular social organization and food specialization, nonhuman primates appear especially skilled at deriving meaning and value from nature. The following review suggests that this may be a derived emotion-mediated process along the lines of the biophilia hypothesis.

Alas, we cannot turn the table on the Saunders and Woods study and ask nonhuman primates how they feel when they encounter their human relatives. Our available methods preclude us from exploring this possible aspect of primate biophilia.[3] Instead, we review encounters among groups of different primate species. We also focus on skills in exploration and exploitation that we feel may be characteristic of the primate relationship with nature. Interpreting existing accounts of primate behavior from a biophilia perspective implies charting new territory, and our approach necessarily is illustrative and anecdotal. Although anecdotes are not data,[4] we nevertheless hope that our annotated review will encourage both primatologists and child developmentalists to develop further systematic investigations of biophilia.

Perceiving Nature

A good start for our review is to address the question "How do primates perceive nature?" Most of what we know involves visual perception, and we know that most primates share basic features of visual perception with us. Like us they have stereoscopic vision, and like us they see the world in color. Chimpanzees, our closest relatives, for instance, have visual acuity that is very similar to ours and perceive colors and shapes in much the same way as we do (Matsuzawa, 1985, 1990, cited in Tomasello & Call, 1997).

There is more to the question of how primates perceive nature than meets the eye, however. Mainstream cognitive psychologists have long believed that nature is "in our head" rather than outside our eyes. This traditional view of perception holds that mental representations of natural environments pass through a system of cognitive schemas before any knowledge of them is obtained. As pointed out by Kahn (1999) and others (Soulé & Lease, 1995), postmodern (especially deconstructivist) thinkers tend to expand on this traditional view by claiming that nature is a mere cultural convention or artifact. From an environmental standpoint, the implications are startling: if nature is no longer natural, there's no reason anymore to attempt to preserve it (see Soulé & Lease, 1995, for responses to postmodern deconstructionist arguments about nature). This general perspective has been challenged by ecological psychologists who dismiss the notion that the meaning of nature is cognitively construed (E. J. Gibson, 1991; cf. J. J. Gibson, 1979; see also Wohlwill, 1983; Reed, 1996). They argue instead that there is no need to postulate mental representations or cognitive schemas: nature is veridical, and we are equipped to directly perceive nature's invariants.[5]

Recent findings from a diverse group of scientists, including mathematicians, neuroscientists, engineers, and statisticians, seem to favor ecological psychologists in this debate. The combined findings suggest that natural scenes—environments undisturbed by evidence of human civilization—exhibit a surprising degree of statistical similarity (Olshausen

& Field, 2000). That is, if we analyze natural scenes at the level of pixel intensity, we find that the distribution of luminance values follows highly predictable patterns. More to the point, this research shows that primate visual perception (human and nonhuman) is exquisitely fine-tuned to pick up these statistical regularities (ibid.).

The adaptive value of such a direct perception system seems obvious. Take, for example, the fast-paced arboreal locomotion that is characteristic of many tree-dwelling primate species. To leap safely from canopy to canopy it is crucial to have the ability to perceive instantly whether the next branch is able to support a landing. Moreover, before leaping the animal has to account for his leaping abilities and present position in space. Or, to put it more formally, the *affordances* (what natural objects furnish, for good or ill) of the arboreal canopy are meaningful only in relation to an individual's action capabilities and specific experience. For primates, directly perceiving nature's affordances thus appears to go hand in hand with actively perceiving the self (Spada, Aureli, Verbeek & de Waal, 1995; see also E. J. Gibson, 1991). This perceptual interplay between self and nature's affordances constitutes the most basic relationship between primates and their natural world.

Encountering Nature, Part I

Exploration

Nonhuman primates encounter a variety of mainly tropical or subtropical habitats, ranging from arid deserts to lush rain forests. To survive and ultimately thrive in a particular habitat, individuals need to be attuned to the meaning and value of habitat-specific affordances. To derive meaning from nature primates actively engage in exploratory behavior. Depending on type of habitat, range size, food specialization, and degree of gregariousness of the species, primates may spend a significant amount of their waking hours on exploration, or "effort after meaning," as ecological psychologist Edward Reed (1996) calls it (cf. Oates, 1987).

Exploratory skill increases with age, and the foundation for effective exploration is most likely laid during early development. Most primates go through an extended period of development, and evidence suggests that juvenile play may exert a special role in becoming attuned to local affordances (Fagen, 1994). Shepard (1995) picks up on this theme and relates it to our own development. He emphasizes the play space—trees, shrubs, paths, places to hide and climb—as a part of the ontogeny for which we have been shaped through evolutionary time but for which there is often no place anymore in modern society. Bernhard (1988) has a similar perspective and elaborates on the role of emotion in play and exploration in both human and nonhuman primates, as we see later in the chapter.

Exploitation

Perceiving meaning is a necessary step toward successfully deriving value (sustenance) from nature. To survive and thrive, "effort after meaning" must be complemented by "effort after value" (Reed, 1996). Field studies show that particular types of learning are involved, albeit not necessarily the multitrial learning of specific responses that captive primates (and their rodent counterparts) so faithfully demonstrate within the limited confounds of behaviorist laboratories. Instead, naturalistic observations show that successful habitat exploitation depends more on rapid learning and behavioral flexibility and less on contingency-based rigid response patterns (Byrne, 1995).

Social Nature

For many primate species effort after meaning and value is facilitated by living in a social group. Two can see more than one, and three more than two, which can come in handy when predators are on the prowl. Moreover, individuals tend to differ in temperament and experience, and a discovery by one individual can potentially benefit another (cf. Kummer, 1971). Novel ways of exploring and exploiting natural affordances can be passed on from individual to individual through social learning mechanisms. For instance, when one individual approaches or contacts something of potential value, chances are that other individuals may do the same. This type of social learning, commonly referred to as *stimulus enhancement*, may be especially effective in enhancing effort after meaning (Byrne, 1995).

Exploitation of the natural environment may be facilitated by *emulation* or *imitation*. In the former case an individual duplicates the results of another group member's actions. In the latter case an individual copies the exact actions of another individual to obtain a similar goal (Byrne, 1995). Emulation may be more prominent than imitation in most primate social groups (Tomasello, 1995), but both types of social learning are probably equally effective in passing on exploitative skills (Whiten et al., 1999).

Of course, sharing one's life with a group of conspecifics has its disadvantages as well. Food may be scarce at times, and competition with other group members may hamper an individual's effort after value. Being attuned to the actions of others may help in gaining a competitive edge, however. Chimpanzees, for instance, have been shown to be especially attuned to what conspecifics can and cannot see and may use this knowledge to their advantage while competing for food (Hare, Call, Agnetta & Tomasello, 2000). Being able to predict the behavior of group members may also help individuals to compete effectively, and monkeys and apes are remarkably skillful at reading and anticipating the behavior of conspecifics (Whiten, 1996; cf. Andrews, 2000).

Emotional Nature

Human developmental research inspired by the influential developmentalist and biologist Jean Piaget portrays childhood as a period during which we are particularly motivated to seek out the natural world around us. Shepard (1983) refers to this childhood process as "loading the ark." Of course, active exploration of the natural world is not confined to our childhood years. In fact, a lifelong propensity to explore and exploit the natural world is most certainly a trait we share with all primate species.

Considering the prominence of this trait, both during development and later in life, the question arises as to what drives us—human and nonhuman primates alike—to actively seek meaning and value from the natural environment? Explanations solely based on instinct or basic feedback systems, such as those associated with hunger and thirst, tend to come up short, although these systems are certainly part of the mix.

Despite its enduring appeal to some, Descartes's famous assertion that human beings "consciously" seek knowledge from nature while animals "automatically" explore and exploit is of little help to us here. For our present level of analysis, Voltaire, one of Descartes's contemporary critics, is a better guide. In response to Descartes's claim that animals are mere automata, Voltaire replied by asking: "Has nature arranged all the means of feeling in this animal, so that it may not feel? Has it nerves in order to be impassable? Do not suppose this impertinent contradiction in nature" (Voltaire, cited in Regan & Singer, 1989).[6]

As Voltaire correctly predicted, we now know that many aspects of animal-environment encounters are accompanied or driven by emotion (McLean, 1952; Panksepp, 1989; Lazarus, 1991; Damasio, 1999). The roots of this system date back to early evolutionary history. The limbic system, the area of the brain underlying the neocortex that mediates emotion, is shared by many animals, including all primate species. And one of its likely main functions is to mediate exploration and exploitation of what nature affords (Crook, 1989; Lott, 1991; Whiten, 1996; Weisfeld, 1997). In primates, the system also plays an additional important role in social interaction (e.g., Maestripieri, Schino, Aureli & Troisi, 1992; van Hooff & Aureli, 1994; Aureli & Smucny, 2000).

A useful heuristic from Bernhard (1988) ties recent empirical and theoretical work on the mediating role of emotion to the construct of biophilia. He suggests that three major emotional systems are intimately connected with both human and nonhuman primate learning—the emotions of attachment, belonging, and security; the emotions of individual identity and status; and the emotions of investigation and discovery.

Much has been written about emotions of attachment, belonging, and security in nonhuman and human primates. For example, most, but not all, available evidence suggests that a youngster's secure attachment to the primary care giver generally allows for the development of effective exploratory and social skills. Conversely, insecure attachment may hamper learning about nature and the social group (for nonhuman primates, see Harlow & Harlow, 1965; for children, see Erikson, Sroufe & Egeland, 1985; see Lamb, Thompson, Gardner & Charnov, 1985). For example, the way young brown capuchin monkeys learn to deal with the aftermath of social conflict depends, in part, on the emotional quality of their relationship with their mother (Weaver & de Waal, 2000).

Similarly, a growing literature deals with emotions of individual identity and status, particularly dominance status, in monkeys and apes (e.g., Sapolsky, 1993, 2000). Both of these emotion systems are relevant to the construct of biophilia as they deal with attraction and fear of living things. Compared to these two systems, less is known about the third system proposed by Bernhard (1988)—the emotions of investigation and discovery—and for the remainder of this chapter we concentrate on this third system and its relevance to biophilia. As Bernhard (1988) explains, for young (human and nonhuman) primates the learning activities of observation, imitation, exploration, and play are pleasurable and exciting, encouraging young primates to learn the exploitative skills that are necessary to survive and thrive (cf. Fagen, 1994). Coles describes this process as it applies to an eight-year-old Hopi Indian girl from the Arizona mesa: "I started realizing how probing a naturalist she is (which is not unusual among Hopi children) and (more extraordinary) how preoccupied she could become: her mind seemed almost lost in thought, so engrossed was she with the land and the sky, the sun, moon, and stars, the flowers her mother grew, the animals, the changes of light that came with clouds" (Coles, 1990, quoted in Nabhan & Trimble, 1994, pp. 125–126). In the following sections we explore how emotion mediates nonhuman primate investigation, discovery, and mastery of nature.

Encountering Nature, Part II

Avoiding Danger

Mastering nature involves both exploiting what nature has to offer and avoiding nature's potential dangers. Poisonous snakes pose a potential threat to human and nonhuman primates alike, and few of us denies feeling at least somewhat uncomfortable when confronted with one of these mysterious elongated creatures. Kanzi, the bonobo, Sue Savage-Rumbaugh's trusted partner in her quest for an understanding of the great ape mind (Savage-Rumbaugh & Lewin, 1994), acts no differently from us when confronted with a snake. Consider, for example, the following incident (Savage-Rumbaugh, Shanker & Taylor, 1998, p. 36):

Just before the Treehouse, I felt Kanzi's body begin to stiffen, and I noticed that the hair on his legs, which was all I could see of him when he was astride my shoulders, was beginning to become erect. Kanzi made a soft "Whuh" sound and gestured to the side of the trail. There, a short distance from my foot, was coiled a large snake.

Kanzi is not alone. A number of studies on different species suggest that primates share the peculiar mix of fear and fascination with snakes that people the world over are apt to show (for a review, see Ulrich, 1993). Why this widespread emotional reaction to snakes? For one, susceptibility to classical or instrumental conditioning may be of little use when faced with the immediate threat of a poisonous snake. An emotional propensity to be alert for the prospect of encountering a snake, allowing for quick evasive action when one crosses one's path, seems far more useful. Applying the laws of natural selection, it seems likely that at some time in the primate past, individuals who were born with a genetically based trepidation for snakes were more likely to survive and reproduce than individuals who lacked such sensitivity. What we need to know more about is how such an innate emotion-based advanced-warning system mediates rapid learning.

Exploration and Exploitation

Kanzi lives at the research facilities of Sue Savage-Rumbaugh and Duane Rumbaugh in Atlanta, Georgia. The research center is surrounded by a 55-acre forest. Kanzi and his human caretakers regularly explore the forest on extended hikes. As Savage-Rumbaugh explains, Kanzi enjoys climbing the trees in the forest and spends much time looking at small animals and insects and learning about the naturally edible plants in the forest (Savage-Rumbaugh & Lewin, 1994; Savage-Rumbaugh et al., 1998). Kanzi quickly learned the various locations in the forest where natural food is available and where food has been hidden by the research staff (ibid.).

Kanzi's skills in exploring and exploiting his Georgia forest reflects the skills of his species in the wild. In their forest home in the heart of equatorial Africa, bonobos travel in small groups from one feeding site to the next. They eat a wide range of plant foods but specialize in ripe fruit (de Waal & Lanting, 1997). Fruit resources may be scarce and are often widely dispersed. Random forays in search of enough ripe fruit could potentially be a waste of energy that wild bonobos can ill afford. Systematic observations of free-ranging bonobos suggest that their travel patterns are instead based on an extensive knowledge of the forest and its fruiting patterns (Kano, 1992).

Although considered to be among the brightest members of the primate taxon, bonobos are by no means exemplary in their mastery of navigating and exploiting their native forest. Field research shows that such skills are widely dispersed among monkeys and apes (for reviews, see Garber, 2000; Tomasello & Call, 1997). For instance, Garber and Hannon (1993) observed the foraging patterns of tamarins, a small neotropical monkey species. Following their observations Garber and Hannon compared their data to three computer models: one represented a random foraging pattern, one was based on olfactory cues, and the third predicted visitations to the closest tree not recently depleted. The analysis showed that the third model best matched the observed foraging patterns. Garber and Hannon concluded that their subjects probably knew the location of hundreds of trees in their home range and remembered for long periods which were still worth a visit and which were depleted.

Menzel (1991, cited in Tomasello & Call, 1997) conducted an ingenious field experiment to investigate "effort after value" in free-ranging Japanese macaques. Menzel conducted his experiment during the time the native akebi fruit, a favorite of Japanese monkeys, was out of season. He placed either a ripe akebi fruit, a piece of chocolate, or nothing at all (the control condition) beside a trail regularly traveled by the monkeys. He waited until a monkey discovered the provisioned food (or approached the area in the control condition) and then followed that monkey for the next 20 minutes. Monkeys who found the ripe akebi fruit immediately stared into the trees where akebi vines might be growing or went to locations where akebi vines grew and inspected them. All of this occurred significantly more often than when either chocolate or nothing was placed along the trail. The monkeys who found chocolate explored the ground near the provisioning site and returned to that site during the 20-minute trials more often than did monkeys in the other two conditions.

Emotional Mediation

What stands out from these observations—and we could have listed many more—is that primates are skilled at exploiting their natural habitat. Much of the research on such mastery has focused on cognitive correlates (e.g., Milton, 1988, 2000; Garber, 2000), however, and there are few systematic investigations of how emotion mediates successful exploration and exploitation. It seems reasonable, however, to expect that emotion plays an equally important role in learning to exploit nature's bounty as it does in learning to avoid nature's dangers. Finding plentiful food, for instance, has been shown to evoke strong emotional reactions and mediate reciprocity in wild chimpanzees. A foraging party that discovers an abundant food source may respond by hooting and drumming on trees, attracting other members of the larger community to a scene described by Reynolds and Reynolds (1965) as "carnival." Goodall (1986) reports that at times a traveling party of chimpanzees may utter anticipatory food calls for up to three minutes before arriving at a known food source, such as a fruiting tree. When females are traveling together, a subordinate individual may touch or embrace a dominant travel companion before rushing toward the desired food (ibid.).

In captivity branches with fresh leaves are ideal to learn about sharing; they arouse quite a bit of excitement yet no excessive competition. When the chimpanzees at the Yerkes Field Station see a caretaker arrive in the distance with two enormous bundles of blackberry, sweetgum, beech, and tulip tree branches, they burst out hooting. General pandemonium ensues, including a flurry of embracing and kissing. Friendly body contact increases 100-fold compared to baseline, and status signals 75-fold (de Waal, 1992). Subordinates approach dominants, particularly the alpha male, to greet them with bows and pant-grunts. Paradoxically, this means that the apes confirm the dominance hierarchy right before canceling it for all intents and purposes. De Waal (1992, p. 37) calls this event a "celebration." It marks the transition to a mode of interaction dominated by tolerance and reciprocity. Celebration serves the elimination of social tensions and thus paves the way for a relaxed feeding session.

Manipulating Nature

Tool Use

Increasing evidence from field studies shows that primates are particularly adept at putting knowledge about their environment to sophisticated use. This evidence suggests that what has evolved is not merely the use of nature's affordances but also the awareness of what things afford and the concomitant ability to select, find, and extract relevant affordances from the environment (Reed, 1996). At Gombe, Goodall's chimpanzee research site of almost 40 years, termites are obtained almost exclusively through the use of tools (Goodall, 1986). To get at the termites inside the termite mound, chimpanzees insert probes fashioned from grasses, vines, bark, or twigs. In defense of their colony the termites cling to the probe, allowing the chimpanzees to extract them from the mound and swallow them with apparent delight. The Gombe chimpanzees prepare the probes by stripping twigs from their leaves or by bending grassy stems, and as such the chimpanzees both use and manufacture tools. Advance planning also appears to be involved because termite-fishing tools are sometimes prepared in anticipation of their use at a mound that is still at some distance and out of sight.

Primate Cultures

Prior to Goodall's discovery of chimpanzee tool making, this sophisticated effort after value was seen as a hallmark of human culture. Since Goodall's initial discovery, reports on different types of tool making and tool use in Gombe and other field research sites have proliferated. Whiten and his coauthors (1999), several of whom have studied wild chimpanzee populations for decades, provide a comprehensive overview of the various chimpanzee tool technologies, ranging from "resin pounding" to "nut-hammering" (cf. McGrew, 1994). Tool making and tool use are learned behaviors that are passed on to the next generation through social learning mechanisms. As Whiten et al. (1999) point out, populations of chimpanzees differ on the use or nonuse of tools as well as on tool technology. Because tool technology is learned rather than innate and because use and nonuse, or variation in tool technology, does not appear to be linked to ecological factors, primatologists have come to refer to population differences in tool technology as differences in culture (de Waal, 1999).

Now that important work on the description, definition, and classification of primate cultural behavior is well under way, it seems to us that an investigation of the role of emotion in cultural behavior has the potential to further our knowledge of biophilia. We expect, for example, that the social transmission of population-specific behavior is mediated by emotion. Detailed data that allow for such an analysis may already be available. For instance, Matsuzawa (1994) obtained detailed records of palm nut hammering by a chimpanzee population in Guinea, West Africa. To crack the palm nuts the chimpanzees commonly use one stone as an anvil and another as a hammer. Individuals within the group tend to reuse their proper set of tools. The chimpanzees in this population do not reach an adult level of skill until they are nine to 10 years old. Youngsters spend a lot of time near adults paying close attention to the adults' progress. Juveniles regularly attempt to replicate their mother's behavior, raising the question of which type of social learning best explains the transmission of this skill to the younger generation—teaching by adults, emulation, imitation, or a combination of these learning mechanisms. A detailed study of the role of emotion in this process may help provide an answer to this question.[7]

Medicinal Plant Use

Another aspect of effort after value that appears to be mediated by emotion is the ability of great apes to learn to use certain plants for medicinal purposes. Researchers report that some of these plants are bitter medicine indeed, but bad taste doesn't seem to keep the apes from consuming them (Huffman & Wrangham, 1994). In the case of these apes, feeling bad appears to override a natural aversion for bad-tasting food that is common in many species. Such an aversion for foul-tasting food is generally adaptive, since such food may be contaminated or inherently toxic and thus cause disease or even death. Both chimpanzees and bononos, as well as Eastern lowland gorillas, have been shown to self-medicate in this fashion. Based on behavior, plant pharmacology, and ethnomedical information, researchers expect that the medicinal values of these plants include the control of parasites, treatment of gastrointestinal disorders, regulation of fertility, and possibly antibacterial or antihepatotoxic activity (ibid.).

Kinship with Nature

So far our review has illustrated that nonhuman primates are keenly attuned to their natural habitat and are masters at deriving value from it. And as the biophilia hypothesis predicts, learning to explore and exploit nature appears to involve emotional mediation at various levels. At this point the question arises as to whether there are additional aspects to the primate relationship with nature. In particular, can we detect elements of kinship with nature? In the next sections we explore this intriguing possibility.

Deriving Pleasure from Nature

Caretakers of captive primates are well aware that exposure to enclosures that feature grass, soil, or other natural features enhances the psychological and physical well-being of their primate charges. Fouts (1997, p. 323), for instance, describes how the chimpanzees in his language-research laboratory reacted to their new grassy outdoor enclosure after having spent many years inside:

> Dar squeezed by and exploded out the door and down the stairs to the ground. He raced across the grass field with such an ecstatic movement that he looked like he was skipping, quadrupedally. . . . Washoe was the next one out. She stood upright and surveyed the terraces, the garden and the familiar human faces at the observation window below. Stretching out her leg, she touched her toes to the first step and pulled them back quickly. Then she noticed Debbi [Fouts] standing at the fence near her. She walked over, reached through the fence and kissed Debbi through the wire.

Perhaps not unlike a city-dweller's deeply felt contentment during a Sunday walk in the park, the reaction of Fouts's chimpanzees to their reintroduction to natural things suggests an emotional attachment to nature that even an extended time away from nature could not erase.

A Sense Of Wonder

In *The Descent of Man* Charles Darwin wrote that he believed that certain animals are capable of a sense of wonder (Darwin, 1871/1981).

There are a few anecdotal accounts that suggest that nonhuman primates may possess such a sense of wonder. In a short paper, Bauer (cited in Konner, 1982) describes how Gombe chimpanzees reacted to encountering the dramatic natural display of a waterfall. Bauer described their reaction as a mixture of silent contemplation and euphoric celebration. Unless important cues were overlooked, it seems that there's no other explanation for this behavior than one of sheer exaltation. Heavy rain seems to trigger behavior that resembles the waterfall displays. During heavy rain most activity is depressed, and the chimpanzees seem intent on sitting out the storm (Goodall, 1986). However, the onset of heavy rain is often marked by wild male displays, often punctuated by aggressive incidents (ibid.). The boisterous male displays have earned the label "rain-dance" and have been observed in other study populations as well (Whiten et al., 1999).

Another observation of an event that may reflect a sense of wonder in chimpanzees was obtained at the fieldstation of the Yerkes Regional Primate Center. As de Waal (1996) describes it, in the middle of a summer day the entire chimpanzee colony gathered around a female called Mai. All were silent and stared at Mai's behind, some poking a finger at it and then smelling the finger. Mai was standing half upright, with her legs slightly apart, holding one hand between her legs. An attentive older female mimicked Mai by cupping her hands between her own legs in the same way. After about 10 minutes, Mai tensed, squatted more deeply, and caught a baby in both hands. The crowd stirred, and Atlanta, Mai's best friend, emerged with a scream, looking around, and embracing a couple of other chimpanzees next to her, one of whom uttered a shrill bark.

De Waal points out that the chimpanzees seemed as much interested in the birth process as in the outcome. The reaction by Mai's best friend, Atlanta, may have been mediated by empathic concern for the plight of her friend. Empathic emotions may also have spread through the group by emotional contagion, a more basic form of empathy.[8] Moreover, perhaps some of the chimpanzees' excitement also stemmed from a genuine sense of wonder associated with witnessing the arrival of a new life in their midst.

During a series of observations of a large captive group of mandrills at a Florida zoo, the first author observed another example of what might be described as an animal sense of wonder. The mandrills usually spent much of their day digging in the soil of their large outdoor enclosure, reminiscent of wild mandrill's explorations for insects, roots, or tubers. Individuals would normally spread out while engaging in this foraging-like behavior, with youngsters staying close their mothers, and the dominant female usually staying close to the large adult male. One afternoon the entire group interrupted their usual digging and rushed over to the adult male who was sitting in the back of the enclosure. The male appeared to touch an object, ever so lightly, that was lying in front of him. Closer inspection through the zoom lens of the observer's video camera revealed that the object was a toad who was not doing a very good job at playing dead. Never before during the months of observations had the entire group clustered tightly around the adult male. And never before had they shown such intense interest for an object, whether inanimate or animate. When the video of the entire event was played back, we noted that the infants and juveniles in the group appeared especially transfixed by the trespassing amphibian. If an early sense of wonder predicts the good naturalist, these mandrill youngsters seemed fit for a career in field biology. Eventually, after careful examination by the monkeys, the toad was let go and scurried off, apparently unharmed by its primate encounter.

Primate Encounters

Ecologists speak of *facilitation* when one population's fitness[9] is increased by the presence of a population of another species. Mutualism occurs when the fitness of two populations of different species sharing a habitat is mutually facilitated. Besides being linked adaptively, populations can also compete for resources or coexist in a shared habitat without much of an effect on each other's fitness. There are various reports, for instance, of habitat sharing in apes that appears neutral in terms of fitness consequences. At the Nouabalé-Ndoki National Park, Congo, gorillas and chimpanzees have been observed cofeeding on figs during the lesser fruiting season (Kuroda, Nishihara, Suzuki & Oko, 1996). During one cofeeding episode, chimpanzees and gorillas made bed sites at distances of less than 50 meters apart, both near the fig tree. Yamagiwa, Maruhashi, Yumoto, and Mwanza (1996) reported from a study conducted as Kahuzi-Biega National Park, Zaire, that when gorillas and chimpanzees met around fruiting trees popular by both populations, they remained calm and seemed to avoid confrontation.

The relationship between sympatric apes and monkeys is decidedly more complex and ranges from mutual attraction to predator-prey relationships. Chimpanzees, for instance, hunt and kill monkeys for their meat. Meat eating is less frequent in bonobos, and their relation with monkeys is far removed from that between predator and prey (de Waal & Lanting, 1997). In Wamba, monkeys have actually been seen to groom and play with bonobos, but there are also reports of forced interactions between apes and monkeys. In such instances bonobos seem to regard monkeys as mere toys (ibid.).

Interactions between monkey species are complex as well, and the pros and cons of such associations in terms of their possible effect on fitness are still relatively poorly understood (Waser, 1987). Explanations of positive aspects of interspecies associations include improved foraging, increased safety from predators, and social benefits, such as receiving grooming (ibid.). Social benefits may include developmental benefits as well. In Uganda, for instance, juveniles of other species commonly play with Colobus monkeys. The rate at which they do this is related to the makeup of their natal group. If there are fewer playmates available in the natal group, juveniles of sympatric species are more likely to seek out Colobus peers (ibid.).

An unstated assumption in these reports is that individuals from different species indeed do recognize each other as fellow primates. In other words, rather than, for instance, soliciting grooming from porcupines (which, of course, would not be prudent), monkeys solicit grooming from other hairy mammals who tend to live in trees, dwell in social groups, and display a wide range of social signals. Are monkeys indeed aware of their kinship with primate relatives? Do they recognize primate features in their next of kin? And if so, which are some of the features they tend to focus on? If it indeed exists, how did such mutual recognition arise? What role does emotion play in interactions between individuals of different species? Finding answers to these questions will help advance our overall understanding of primate biophilia.

Conclusions

We have come full circle. We started our presentation with thoughts about our human fascination with primate kin. We ended it with accounts of mutual attraction between different primate species. Throughout our review we looked at how primates perceive, explore, and exploit nature and presented anecdotes that are suggestive of primate kinship with nature. We attempted to build a case for primate biophilia by focusing on direct encounters with natural things. We have taken our cues, in part, from ecological psychology and behavioral biology; work in both fields clearly suggests that to understand nature's psychological pull on primates, we should not look for something in the animal but rather look at the animal in its world (cf. Gibson, 1994; Reed, 1996). Our approach illustrated that primates are keenly attuned to their natural environment and are skilled at exploiting what nature affords. Although systematic studies of primate emotions in natural settings are still few and far between, we believe that our selected accounts suggest that learning from and about nature may involve emotional mediation at multiple levels. In fact, as the biophilia hypothesis predicts, our review suggests that relating to nature is not a cold-blooded affair for monkeys and apes.

How does primate biophilia relate to children's early relationship with nature? The contemporary developmental niche[10] of human children certainly differs in many ways from that of the young of both our early human ancestors and our closest living relatives, the nonhuman primates, particularly with respect to the complex sociality that defines our species today. As Sinha (1985, p. 161) explains: "The child develops by expanding its known habitat through exploration and this widening circle of adaptation, based on direct perception of the environment, includes and necessitates the support of the social structure as it has been meaningfully shaped by man." However, some of the basic ingredients of this early developmental process are undoubtedly shared among the young of all primates. After all, extant primates and human beings have walked the evolutionary road together for a long time and, along the way, faced similar problems in finding their way in nature. In particular, direct perception and the way that the emotions of investigation and discovery mediate learning about and from nature are psychological particularities that we feel are most likely shared by the young of human and nonhuman primates alike.

Notes

1. *Attunement* refers to an individual's ability to detect the various information structures in the environment.

2. Prototaxis, the innate tendency of one organism or cell to react in a definite manner to another organism, may be another component part of biophilia. A detailed discussion of this and other possible precursors of biophilia is beyond the scope of this chapter.

3. In a series of elegant experiments with chimpanzees, Sarah Boysen and her colleagues showed that heart rate was correlated with visual recognition of both humans and conspecifics (reviewed in Boysen, 1994). Perhaps this methodology can be adapted to measure physiological correlates of cross-species biophilia in primates.

4. Credit goes to primatologist Irwin Bernstein for this bit of wisdom.

5. Ethologist Konrad Lorenz expressed a similar view in his *Die Rückseite des Spiegels*. He argued that the ability of animals to pick up natural invariants enhances their interactions with the natural environment (Lorenz, 1973, cited in Charlesworth, 1978).

6. Thanks go to Kristin Andrews, who pointed out the Voltaire quote to the first author and who cited it in her doctoral dissertation on belief attribution in human and nonhuman primates (Andrews, 2000).

7. Ottani and Mannu (in press) observed remarkably similar nut-hammering behavior in semi-free-ranging brown capuchin monkeys.

8. Emotional contagion is "the tendency to automatically mimic and synchronize expressions, vocalizations, postures, and movements with those of another person and, consequently, to converge emotionally" (Hartfield, Cacioppo & Rapson, 1993).

9. *Fitness* in this context refers to survival and reproductive success.

10. A developmental niche is "the physical and social context in which the child lives, including child rearing and educational practices, as well as psychological characteristics of the child's parents" (cf. Super & Harkness, 1986, p. 545).

References

Andrews, K. (2000). Predicting mind: The role of belief attribution in philosophy and psychology. Doctoral dissertation, Department of Philosophy, University of Minnesota.

Atran, S. (1990). *The cognitive foundations of natural history.* New York: Cambridge University Press.

Aureli, F., & Smucny, D. (2000). The role of emotion in conflict and conflict resolution. In F. Aureli & F. B. M. de Waal (Eds.), *Natural conflict resolution.* Berkeley: University of California Press.

Barkow, J. H., Cosmides, L., & Tooby, J. (1992). *The adapted mind.* New York: Oxford University Press.

Bernhard, J. G. (1988). *Primates in the classroom.* Amherst: University of Massachusetts Press.

Boysen, S. T. (1994). Individual differences in the cognitive abilities of chimpanzees. In R. W. Wrangham, W. C. McGrew, F. B. M. de Waal & P. G. Heltne (Eds.), *Chimpanzee cultures.* Cambridge, MA: Harvard University Press.

Byrne, R. (1995). *The thinking ape.* New York: Oxford University Press.

Charlesworth, W. R. (1978). Ethology: Understanding the other half of intelligence. *Social Science Information, 17,* 231–277.

Coles, R. (1990). *The spiritual life of children.* Boston: Houghton Mifflin.

Crook, J. H. (1989). Introduction: Socioecological paradigms, evolution and history—Perspectives for the 1990s. In V. Standen & R. Foley (Eds.), *Comparative socioecology: The behavioral ecology of humans and other mammals* (pp. 1–36). Oxford: Blackwell Scientific.

Damasio, A. R. (1999). *The feeling of what happens.* New York: Harcourt Brace.

Darwin, C. R. (1981). *The descent of man, and selection in relation to sex.* Princeton, NJ: Princeton University Press. (Original Work Published 1871.)

de Waal, F. B. M. (1992). Appeasement, celebration, and food sharing in the two *Pan* species. In T. Nishida, W. C. McGrew, P. Marler, M. Pickford & F. B. M. de Waal (Eds.), *Topics in primatology,* Vol. 1, *Human origins.* Tokyo: University of Tokyo Press.

de Waal, F. B. M. (1996). *Good natured: The origins of right and wrong in humans and other animals.* Cambridge, MA: Harvard University Press.

de Waal, F. B. M. (1999). Cultural primatology comes of age. *Nature, 399,* 635–636.

de Waal, F. B. M., & Lanting, F. (1997). *Bonobo: The forgotten ape.* Berkeley: University of California Press.

Erikson, M. F., Sroufe, L. A., & Egeland, B. (1985). The relationship between the quality of attachment and behavior problems in a high-risk sample. *Monographs of the Society for Research in Child Development, 50* (1–2, Serial No. 209).

Fagen, R. (1994). Applause for Aurora: Sociobiological considerations on exploration and play. In H. Keller, B. Henderson & K. Schneider (Eds.), *Curiosity and exploration.* New York: Springer.

Fouts, R. (1997). *Next of kin*. London: Joseph.

Garber, P. A. (2000). Evidence for the use of spatial, temporal, and social information by some primate foragers. In S. Boinski & P. A. Garber (Eds.), *On the move*. Chicago: University of Chicago Press.

Garber, P. A., & Hannon, B. (1993). Modeling monkeys: A comparison of computer-generated and naturally occuring foraging patterns in two species of neotropical primates. *International Journal of Primatology, 14*, 827–852.

Gelman, R. (1990). First principles organize attention to and learning about relevant data: Number and the animate-inanimate distinction as examples. *Cognitive Science, 14*, 79–106.

Gelman, R., & Markman, E. (1987). Young children's inductions from natural kinds: The role of categories and appearances. *Child Development, 58*, 1532–1540.

Gibson, E. J. (1991). The ecological approach: A foundation for environmental psychology. In R. M. Downs, L. S. Liben & D. S. Palermo (Eds.), *Visions of aesthetics, the environment, and development: The legacy of Joachim F. Wohlwill* (pp. 87–112). Hillsdale, NJ: Erlbaum.

Gibson, E. J. (1994). Has psychology a future? *Psychological Science, 5*, 69–76.

Gibson, J. J. (1979). *The ecological approach to visual perception*. Boston: Houghton Mifflin.

Goodall, J. (1986). *The chimpanzees of Gombe: Patterns of behavior*. Cambridge, MA: Belknap Press of Harvard University Press.

Haraway, D. J. (1991). *Simians, cyborgs, and women: The reinvention of nature*. New York: Routledge.

Hare, B., Call, J., Agnetta, B., & Tomasello, M. (2000). Chimpanzees know what conspecifics do and do not see. *Animal Behaviour, 59*, 771–785.

Harlow, H. F., & Harlow, M. K. (1965). The affectional systems. In A. M. Schier, H. F. Harlow & F. Stollnitz (Eds.), *Behavior of nonhuman primates* (vol. 2). London: Academic Press.

Hartfield, E., Cacioppo, J. T., & Rapson, R. L. (1993). Emotional contagion. *Current Directions in Psychological Science, 2*, 96–99.

Huffman, M. A., & Wrangham, R. W. (1994). Diversity of medicinal plant use by chimpanzees in the wild. In R. W. Wrangham, W. C. McGrew, F. B. M. de Waal & P. G. Heltne (Eds.), *Chimpanzee cultures*. Cambridge, MA: Harvard University Press.

Kahn, P. H., Jr. (1997). Developmental psychology and the biophilia hypothesis: Children's affiliation with nature. *Developmental Review, 17*, 1–61.

Kahn, P. H., Jr. (1999). *The human relationship with nature*. Cambridge, MA: MIT Press.

Kano, T. (1992). *The last ape: Pygmy chimpanzee behavior and ecology.* Stanford: Stanford University Press.

Kellert, S. R. (1996). *The value of life.* Washington, DC: Island Press.

Kellert, S. R. (1997). *Kinship to mastery.* Washington, DC: Island Press.

Kellert, S. R., & Wilson, E. O. (1993). *The biophilia hypothesis.* Washington, DC: Island Press.

Konner, M. (1982). *The tangled wing.* New York: Holt.

Kummer, H. (1971). *Primate societies.* Chicago: Aldine.

Kuroda, S., Nishihara, T., Suzuki, S., & Oko, R. A. (1996). Sympatric chimpanzees and gorillas in the Ndoki Forest, Congo. In W. C. Mcgrew, L. F. Marchant & T. Nishida (Eds.), *Great ape societies.* New York: Cambridge University Press.

Lamb, M. E., Thompson, R. A., Gardner, W., & Charnov, E. L. (1985). *Infant-mother attachment: The origins and developmental significance of individual differences in strange situation behavior.* Hillsdale, NJ: Erlbaum.

Lazarus, R. S. (1991). *Emotion and adaptation.* New York: Oxford University Press.

LeDoux, J. E. (1996). *The emotional brain.* New York: Simon and Schuster.

Lorenz, K. (1973). *Die Rückseite des Spiegels.* Munich: Piper.

Lott, D. F. (1991). *Intraspecific variation in the social system of wild vetebrates.* Cambridge: Cambridge University Press.

Lovelock, J. (1991). Mother earth: Myth or science? In C. Barlow (Ed.), *From Gaia to selfish genes.* Cambridge, MA: MIT Press.

Maestripieri, D., Schino, G., Aureli, F., & Troisi, A. (1992). A modest proposal: Displacement activities as indicators of emotions in primates. *Animal Behavior, 44*, 967–979.

Matsuzawa, T. (1985). Color naming and classification in a chimpanzee (*Pan troglodytes*). *Journal of Human Evolution, 14*, 283–291.

Matsuzawa, T. (1990). Form perception and visual acuity in a chimpanzee. *Folia Primatologica, 55*, 24–32.

Matsuzawa, T. (1994). Field experiments on use of stone tools by chimpanzees in the wild. In R. W. Wrangham, W. C. McGrew, F. B. M. de Waal & P. G. Heltne (Eds.), *Chimpanzee cultures.* Cambridge, MA: Harvard University Press.

McGrew, W. C. (1994). Tools compared. In R. W. Wrangham, W. C. McGrew, F. B. M. de Waal & P. G. Heltne (Eds.), *Chimpanzee cultures.* Cambridge, MA: Harvard University Press.

McLean. P. D. (1952). Some psychiatric implications of physiological studies on frontotemporal portion of the limbic system (visceral brain). *Electroencephalography and Clinical Neurophysiology, 4*, 407–418.

Menzel, C. R. (1991). Cognitive aspects of foraging in Japanese monkeys. *Animal Behavior, 41*, 397–402.

Milton, K. (1988). Foraging behaviour and the evolution of intellect in monkeys, apes and humans. In R. W. Byrne & A. Whiten (Eds.), *Machiavellian intelligence* (pp. 285–305). Oxford: Clarendon Press.

Milton, K. (2000). Quo vadis? Tactics of food search and group movement in primates and other animals. In S. Boinski & P. A. Garber (Eds.), *On the move*. Chicago: University of Chicago Press.

Nabhan, G. P., & Trimble, S. (1994). *The geography of childhood*. Boston: Beacon Press.

Oates, J. F. (1987). Food distribution and foraging behavior. In B. B. Smuts, D. L. Cheney, R. M. Seyfarth, R. W. Wrangham & T. T. Struhsaker (Eds.), *Primate societies*. Chicago: University of Chicago Press.

Olshausen, B. A., & Field, D. J. (2000). Vision and the coding of natural images. *American Scientist, 88*(3), 238–245.

Ottani, E. B., & Mannu, M. (In press). Semi-free ranging tufted capuchin monkeys (*Cebus apella)* spontaneously use tools to crack open nuts. *International Journal of Primatology*.

Panksepp, J. (1989). The psychobiology of emotions: The animal side of human feelings. In G. Gainotti & C. Caltagirone (Eds.), *Emotions and the dual brain* (pp. 31–55). Berlin: Springer.

Reed, E. S. (1996). *Encountering the world*. New York: Oxford University Press.

Regan, T., & Singer, P. (Eds.). (1989). *Animal rights and human obligations*. Englewood Cliffs, NJ: Prentice-Hall.

Reynolds, V., & Reynolds, F. (1965). Chimpanzees of the Budongo Forest. In I. deVore (Ed.), *Primate behavior* (pp. 368–424). New York: Holt, Rinehart & Winston.

Sapolsky, R. (1993). The physiology of dominance in stable versus unstable social hierarchies. In W. Mason & S. Mendoza (Eds.), *Primate social conflict*. New York: State University of New York Press.

Sapolsky, R. (2000). Physiological correlates of individual dominance style. In F. Aureli & F. B. M. de Waal (Eds.), *Natural conflict resolution*. Berkeley: University of California Press.

Saunders, C. D., & Wood, L. (1992). Exploring concepts related to primates and their conservation through focus groups. Manuscript, Communications Research, Brookfield Zoo, Brookfield, IL.

Savage-Rumbaugh, S., & Lewin, R. (1994). *Kanzi: The ape at the brink of the human mind*. New York: Wiley.

Savage-Rumbaugh, S., Shanker, S. G., & Taylor, T. J. (1998). *Apes, language, and the human mind*. New York: Oxford University Press.

Shepard, P. (1983). The ark of the mind. *Parabola, 8*(2), 54–59.

Shepard, P. (1995). Nature and madness. In T. Roszak, M. E. Gomes & A. D. Kanner (Eds.), *Ecopsychology*. San Francisco: Sierra Club Books.

Sinha, C. (1985). A socio-naturalistic approach to human development. In G. Butterworth, J. Rutkowska & M. Scaife (Eds.), *Evolution and developmental psychology*. Brighton, Sussex: Harvester Press.

Soulé, M. E., & Lease, G. (1995). *Reinventing nature? Responses to postmodern deconstruction*. Washington, DC: Island Press.

Spada, E. C., Aureli, F., Verbeek, P., & de Waal, F. B. M. (1995). The self as reference point: Can animals do without it? In P. Rochat (Ed.), *The self in infancy: Theory and research*. Amsterdam: Elsevier.

Super, C., & Harkness, S. (1986). The developmental niche: A conceptualization at the interface of society and the individual. *International Journal of Behavioral Development, 9*, 545–570.

Tomasello, M. (1995). Cultural transmission in the tool use and communicatory signalling of chimpanzees? In S. T. Parker & K. R. Gibson (Eds.), *Language and intelligence in monkeys and apes*. Cambridge: Cambridge University Press.

Tomasello, M., & Call, J. (1997). *Primate cognition*. New York: Oxford University Press.

Ulrich, R. S. (1993). Biophilia, biophobia, and natural landscapes. In S. R. Kellert & E. O. Wilson (Eds.), *The biophilia hypothesis*. Washington, DC: Island Press.

van Hooff, J., & Aureli, F. (1994). Social homeostasis and the regulation of emotion. In S. H. M. van Goozen, N. E. van de Poll & J. A. Sergeant (Eds.), *Emotions: Essays on emotion theory*. Hillsdale, NJ: Erlbaum.

van Lawick-Goodall, J. (1971). *In the shadow of man*. Boston: Houghton Mifflin.

Waser, P. M. (1987). Interactions among primate species. In B. B. Smuts, D. L. Cheney, R. M. Seyfarth, R. W. Wrangham & T. T. Struhsaker (Eds.), *Primate societies*. Chicago: University of Chicago Press.

Weaver, A. C., & de Waal, F. B. M. (2000). The development of reconciliation in brown capuchins. In F. Aureli & F. B. M. de Waal (Eds.), *Natural conflict resolution*. Berkeley: University of California Press.

Weisfeld, G. E. (1997). Research on emotions and future developments in human ethology. In A. Schmitt (Ed.), *New aspects of human ethology*. New York: Plenum Press.

Whiten, A. (1996). When does smart behaviour-reading become mind-reading? In P. Carruthers & P. K. Smith (Eds.), *Theories of theories of mind* (pp. 277–292). Cambridge: Cambridge University Press.

Whiten, A., Goodall, J., McGrew, W. C., Nishida, T., Reynolds, V., Sugiyama, Y., Tutin, C. E. G., Wrangham, R. W., & Boesch, C. (1999). Cultures in chimpanzees. *Nature, 399*, 682–685.

Wilson, E. O. (1984). *Biophilia*. Cambridge, MA: Harvard University Press.

Wilson, E. O. (1993). Biophilia and the conservation ethic. In S. R. Kellert & E. O. Wilson (Eds.), *The biophilia hypothesis.* Washington, DC: Island Press.

Wohlwill, J. F. (1983). The concept of nature. A psychologist's view. In I. Altman & J. F. Wohlwill (Eds.), *Behavior and the natural environment.* New York: Plenum Press.

Yamagiwa, J., Maruhashi, T., Yumoto, T., & Mwanza, N. (1996). Dietary and ranging overlap in sympatric gorillas and chimpanzees in Kahuzi-Biega National Park, Zaire. In W. C. McGrew, L. F. Marchant & T. Nishida (Eds.), *Great ape societies.* New York: Cambridge University Press.

2

The Ecological World of Children

Judith H. Heerwagen and Gordon H. Orians

The developmental psychological literature focuses primarily on how changes in childhood are influenced by the combination of children's experience and the sociocultural environment. However, a full understanding of these changes also requires consideration of the physical and biological environments encountered by children and of our evolutionary history. Thus, in this chapter, our goal is to enrich the dialogue around children and nature to include the perspective of the child's ecological world and to show how conditions experienced in ancestral environments tug at us, sometimes quite strongly, even today.

One of the foundational ideas of this ecological-evolutionary perspective is that current neural capacities and response patterns have evolved as a result of the individuals' past responses to environmental contingencies. The payoff of various responses is determined by statistical association between the perception of different types of information and the consequences of responding in different ways to that information. Responses that contributed positively to fitness were incorporated into the neural machinery; those that reduced fitness were eliminated.

Accordingly, we argue that children's environmental behaviors should show evidence of specific adaptations to enduring challenges and opportunities, from birth through reproductive age. More specifically, in this chapter we predict age-related patterns of behavioral responses. Whenever possible, we test our predictions with published research data. We realize that in doing so we have drawn initially on existing data rather than conducted new experiments with appropriate manipulations and controls. In this way, we may at times be open to the charge that we have made "predictions" to concur with our prior knowledge, although we have tried to be as neutral as possible. Nonetheless, other times we make genuine predictions, some of which are not supported by existing data. When this situation occurs, we suggest research that could put our prediction to the test. Our general argument is based on the idea that environmental information is not equally useful across a human life span. Therefore, what is salient, what is ignored, and how a child responds are expected to vary across ages. Although we limit our attention to childhood, such variable responses presumably continue throughout human life.

The Ecological Context

For making predictions about attention to and responses to environmental contingencies at different ages, it is useful to have a classification of types of environmental information. Information emanating from the physical environment—water, topography, weather, and fire—is relevant to decisions at varied spatial and temporal scales. For example, large-scale topographic information—which, in combination with the gross structure of vegetation, characterizes the landscape in which individuals move—may influence dispersal and selection of a place to live and raise a family. Small-scale topographic information may be critical for deciding the next steps an individual might make. Weather information may signal both seasonal changes and conditions likely to occur in the immediate future.

In addition to physical topography, the structure provided by plants, especially woody individuals, strongly influences the environment at both landscape and more intimate levels. For this reason, we consider vegetation, as opposed to individual plants, as a distinct component of the biological environment.

The remaining elements of the biological environment may usefully be divided into conspecific and heterospecific individuals. For highly social species like humans, familiar and strange conspecific individuals differ strikingly in their significance. Among familiar individuals, a priori expectations typically differ dramatically, depending on whether the individual is a close relative, a competitor, or an individual of very different social status or rank. We use these differences and the ages at which they are likely to be most important as a basis for making predictions about capabilities and responses of children of different ages.

The Physiological and Social Contexts

The genetic constitution of an organism is the result of the past actions of evolutionary agents—that is, all organisms are adapted to past environments. Thus, predictions about patterns of human ontogenetic development should be based on assumptions about the social and ecological worlds in which children were born and in which they matured over the broad course of human evolution.

Hunter-Gatherer Social Organization

Since its origins from a *Homo habilis* type of ancestor about 1.6 million years ago, *Homo sapiens* lived in small groups until as recently as 35,000 years ago. Archaeological evidence of seasonal huts and hearths dates to about 350,000 years ago. Domestication of animals and planting of cereal grains did not begin until about 10,000 years ago. Humans lived in small hunter-gatherer (HG) groups over most of the history of our species, most of which took place in tropical regions. Therefore, we base our predictions about which behavioral traits might have been favored at different maturational stages on the probable conditions encountered by infants and children in HG societies.

It is difficult to determine precisely many components of early HG social organization because most existing HG cultures have been in contact with adjacent agricultural societies for hundreds to thousands of years, during which time they have adopted components of the organization of those societies. Nonetheless, most anthropologists agree that the following features characterized most HG societies prior to the domestication of plants and animals.

Human social groups were small; many members were close genetic relatives. The people had few material possessions that they took with them as they moved their campsite, which they did annually to exploit seasonal changes in food availability in different habitats. Infants were breast fed for at least three to four years; consequently birth intervals were long. Infants were in almost constant physical contact with adults during the first several years of their lives and were carried by their mothers or other adults during foraging forays. Social behaviors were built around reciprocal help in times of danger, food sharing, communal nurturing of all children in the band, and open sharing of knowledge and skills (Lee & DeVore, 1976). The security provided by this social environment has been essential for children's learning throughout human history. This social environment provided opportunities for an extended period of play and exploration. As members of a generalist species, humans need to learn a wide range of behaviors appropriate to foraging and food preparation, as well as avoidance of hazards, both animate and inanimate.

Because social groups were small, the few children present in them would have spanned a broad range of ages. Children would have interacted primarily in mixed-age groups and would have learned much from contacts with older children. Infanticide was probably common as a way of allocating scarce resources or eliminating deformed or sickly infants (Hill & Hurtado, 1996). Intertribal warfare may have been common, but this is less certain.

Peculiarities of Human Development

Human development is remarkably slow for a primate. Humans take 18 to 20 years instead of the 11 years that chimpanzees take to reach full size. Sexual maturity is not achieved until 13 years of age (six to seven in chimps), and human life expectancy is twice that of chimpanzees. The human brain is two to three standard deviations above the line predicting brain size from body size for primates (Pagel & Harvey, 1989). The human brain grows rapidly after birth for about a year, reaching adult size in about seven years (Bogin, 1991).

Human infants are unusually large for a primate of our size. For example, newborn humans average about 3,300 grams, compared with the 1,900 grams of newborn gorillas. A gorilla's pregnancy is barely detectable to the naked eye. Human infants are also remarkably fat, and they continue to lay down fat for several months after birth. As a percentage of its body weight, a human infant has five times more fat than a typical newborn primate.

But although they are large, human infants are basically helpless at birth. An infant gorilla or chimpanzee only two weeks old can climb up to its mother's teat and cling to her fur unaided while she forages. In contrast, a human infant cannot climb, cling, or follow its mother. It cannot even lift its head. Human infants require almost continuous care for the first two years. Therefore, attachment to and bonding with its mother and other caring adults are essential for infant survival (Bowlby, 1980). Not surprisingly, a rich literature exists about the powerful role of early attachment on subsequent human development.

Except at birth, relationships of children to their environments typically change gradually from infancy to adolescence. The emergence of new physical, cognitive, and social skills supports the child's expanding exploration and use of the physical environment. The increasingly sophisticated skills provide the child with an expanding array of behavioral choices. The developmental literature focuses largely on how individual experience and the sociocultural environment combine to influence the appearance of environmentally relevant behaviors. The influence of evolved adaptations is largely ignored. Nonetheless, an adaptive perspective predicts that specific responses to some types of environmental information should appear fairly suddenly and without prior experience. Further, it also predicts that "relevant" information changes over time as children's mobility increases and their vulnerability to hazards decreases.

Predictions About Children's Behavioral Responses to Environmental Stimuli

We order our discussion around the most important types of ecological challenges—safety, foraging and feeding, and finding a place to live. Where possible we discuss challenges in order of the age at which they first become relevant. (We base our predictions on the environmental challenges children have faced throughout human history.) Our predictions about onset and termination of behaviors are based on hypotheses about the ages at which those challenges would have been most important. Table 2.1 summarizes the challenges and predictions. The

Table 2.1
Predictions About Responses to Ecological Challenges

Challenge	When It Becomes Important	Prediction
Safety:		
Staying close to sources of help	When independent movement is possible and individuals can be discriminated	Preferential attraction to small objects close at hand rather than to large objects at a distance
Avoiding hostile conspecifics	When independent movement is possible and individuals can be discriminated	Fear evoked more strongly by strange males than by strange females
Avoiding topographic and climatic hazards	When independent movement is possible but still limited	Fears evoked by sudden changes in sensory conditions; should continue through all ages
Finding refuge	When physical skills are more developed, allowing movement away from caretaker	Fear of small animals (snakes, spiders, insects) developed at young age before fear of large animals
Avoiding large predators	When physical skills are more developed, enabling movement away from caretaker	Attraction to small-scale places that afford protection or shelter Expect fears of large predators to develop later than fear of small animals
Constructing shelter	When physical skills are more developed, enabling greater movement away from home base	Interest in constructing shelter and building activities
Finding the way home	When physical skills are more developed, enabling greater movement away from home base	Fears of being lost
Foraging and feeding:		
Distinguishing edibles from inedibles	When physical skills are more developed, allowing movement away from caretaker	Early mouthing of many objects; termination at weaning
Finding a place to live	At puberty or soon after	Development of interest in landscapes; development of preference for savannahs

subsequent text gives our rationales for the predictions and describes the tests of these predictions that we have been able to conduct.

Throughout their life span, children face numerous risks and hazards generated by their environment. However, the specific threats and hazards they face change with age as children become more mobile and more capable, physically and cognitively, of moving about on their own. Although numerous authors have recognized age-related differences in children's responses to the environment, there has been little attempt to ask why the changes occur and what adaptive function they might serve. Very few researchers, with the exception of Bowlby (1980), have attempted to relate children's environmental responses to their actual use of the environment and to the dangers, opportunities, and uncertainties that they face as they mature and prepare to leave the natal base. Most research focuses on proximate mechanisms such as personal experience or modeling of behavior without any discussion of the physical and biological environment in which the fears occur (Davey, 1995; Muris et al., 1997; Ollendick et al., 1985).

The neural processes that guided our ancestors' behaviors in Pleistocene hunting and gathering bands are likely to still be in operation today (Pinker, 1997; Cosmides & Tooby, 1993). These mechanisms have been designed by evolution to guide adaptive response to enduring ecological challenges—such as distinguishing edible from inedible foods, avoiding encounters with dangerous animals, avoiding dangerous conspecifics, finding the way home, avoiding inanimate hazards, and finding a place to live. The ecological-developmental perspective we present here is based on the idea that environmental information is not equally useful at all ages. Therefore, what is salient, what is ignored, and how a child responds is expected to vary with age, or, more specifically, with developmental abilities.

Predictions Related to Safety

Given the high vulnerability of young children to physical and biological hazards, surviving to reproductive age is a major challenge. Thus, mechanisms and responses for avoiding dangers that could produce serious injuries or death should have evolved. For most species, including humans, key ecological dangers are predators and inanimate hazards (storms and geo-morphological conditions). However, exposure to these hazards is not equally distributed across childhood. As children become more mobile, they encounter new spaces, organisms, objects, and situations that offer both opportunities and dangers. Child mobility also reduces the ease and rapidity with which an adult can come to a child's aid. Thus, as they age, children have to increasingly rely on their own behaviors and can count less on protection and help from adult caretakers. Cries or distress calls work well when a caretaker is within earshot, but not when the child ventures beyond this point. In fact, crying may serve to attract predators and increase the danger to the child rather than reduce it.

A key task faced by a mobile child is to distinguish potentially dangerous features and attributes of situations and stimuli, to assess their implications for well-being, and to act in ways that reduce the potential for harm. Because responses may need to be rapid, we expect much of the processing to be done outside of conscious awareness, generated by programs molded over the long course of human interactions with natural environments as hunter-gatherers.

Fear is a mechanism for avoiding or coping with dangers in many animal species (Archer, 1979; Russell, 1979). King et al. (1988) define fear as a "normal reaction" to a real or imagined situation that is perceived as a threat to one's safety or well-being. (In contrast, phobias are fears that are out of proportion to the situation, cannot be reasoned away, and are beyond voluntary control.)

In the following sections, we elaborate on predictions about childhood fears associated with specific ecological challenges. Although we concentrate on fear-response components that have been programmed by evolution we are fully aware that the social environment and learning exert strong influences on what people fear. Nevertheless, considerable experimental evidence suggest that fear responses may have a genetic basis. The most powerful evidence comes from research with human twins on animal phobias (Kendler et al., 1992) and fear of open spaces (Moran & Andrews, 1985).

Environmental psychologists, particularly in Sweden and Norway, have carried out a series of imaginative experiments on the acquisition and retention of fears to "natural" fear-evoking objects (snakes, spiders), culturally generated fearful objects (handguns, frayed electrical wires), and neutral stimuli (geometric patterns). The general finding of these experiments is that conditioned responses to natural fear-evoking objects are usually, but not always, acquired more quickly and that responses to snakes and spiders are always more resistant to extinction than responses to neutral stimuli (McNally, 1987). Conditioned responses to the modern dangerous stimuli extinguish more quickly when no longer reinforced than do responses to snakes and spiders (Cook, Hodes & Lang, 1986; Hugdahl & Karker, 1981). In addition, aversive responses to fear-relevant natural stimuli can be acquired merely by telling a person that a shock will be administered. Aversive responses to fear-irrelevant stimuli cannot be elicited in this manner (Hugdahl, 1978). These results indicate that fear responses may be genetically influenced.

Infants are born with some defense mechanisms. They clearly cry or show other signs of distress in response to loud sounds, bright light, rapid or irregular movement of stimuli, looming objects, and loss of equilibrium (Smith, 1979; Bower, 1974; Jersild & Holmes, 1935). However, these responses are more reflexive and less like the fear behaviors of older children. Because infants are highly vulnerable when separated from their mother, selection pressures for being able to cope with an array of hazards should be especially strong from the age of six to seven months when crawling behaviors develop. We begin the section on safety with predictions about mechanisms that keep children close to home and to caretakers.

Staying Close to Friendly Conspecifics John Bowlby (1980) has written extensively on mother-infant attachment and the psychological mechanisms that keep an infant safe during early exploration of the environment. However, his work does not consider how an infant's interactions with the environment may influence where and when it moves. Bowlby's theory centers around how interactions between infants and caretakers enable the infant to feel safe as it explores.

These studies show that infants' exploration is centered on small objects. This behavior may serve a survival function that has not previously been recognized. We propose that infants' attraction to small objects evolved in part because it reduces their tendency to wander. If attraction to small objects does help to keep infants close to home, we predict, as a corollary, that infants will ignore larger objects in the distance, even though they are able to see them, in preference for attending to smaller items close at hand. Safety probably is not the primary reason for infants' attraction to small objects, but safety has not been suggested previously as a value of it.

Young children's preference for small, manipulable objects is well known (Garvey, 1990; Chase, 1992). It is, in fact, the defining feature of play for the first two years of life. Active object seeking and exploration begin at about seven months. Studies show that infants and toddlers devote a significant percentage of their waking hours looking at and playing with objects. Clarke-Stewart (1973) found that infants between nine and 18 months spent more time interacting with physical objects (46 percent) than with their mothers (36 percent). Between the ages of 12 and 33 months, children spend 80 to 90 percent of their waking time interacting with objects in the physical environment and only 10 to 20 percent in social interactions (White, Kaban, Shapiro & Attanucci, 1977).

This pattern cannot be explained as a result of inability to perceive distant objects. Studies also show that infant vision is well developed by six to eight months (Berk, 1997). At this age, infants have good visual acuity (20/100 compared to 20/600 at birth), with acuity reaching 20/20 by age two. Thus, they have the ability to see things in the distance. However, do they ignore distant, interesting things in favor of objects close at hand, as we have predicted?

An observational study of mothers and toddlers in an outdoor park (Anderson, 1972) found that the children moved in short bouts of about 9 seconds and then stopped to pick up things at their feet or to sit down and play. The "found" objects included sticks, grass, paper, and stones. The children rarely ventured more than 200 feet from their mothers. The toddlers noticed many distant objects (cars, birds, planes) as indicated by their pointing behavior and vocalizations, but they made no attempt to move toward these objects. Anderson (1972) interpreted pointing as an opportunity for learning and for alerting the mother to potential danger before it becomes an emergency. This interpretation is compatible with ours.

The timing of onset and termination of attractions and aversive responses to some stimuli appears to be more precise than would be expected from a simple developmental perspective. Mouthing of objects develops early and generally terminates at about the age infants would have been weaned during most of human history. Further, it is evident that children have a strong fascination for small manipulable objects but generally show no interest in larger landscape features even though they are capable of perceiving them. Thus, these behaviors are similar to language learning, the capacity for which develops remarkably rapidly and shows that infants innately recognize some underlying rules of grammar (Chomsky, 1975; Pinker, 1997).

Avoiding Hostile Conspecifics We predict that fear of strangers should develop as soon as independent movement is possible. This prediction is supported by the fact that fear of strangers typically begins to develop around seven months, peaks at about one year, and continues until 18 to 24 months of age. Fear of strangers does not develop until several months after discrimination of familiar and unfamiliar persons is possible (Bronson, 1972). The reason for the delay may be that discrimination is only one of the important preconditions for the development of fear of strangers. The other critical developmental milestone is independent movement. Crawling increases the potential for contact with strangers, and it increases infants' vulnerability because they are out of the immediate control of a caretaker. Our suggestion that both discrimination and independent movement are necessary conditions for fear of strangers may explain the developmental "gap" (Zegans & Zegans, 1972) that exists between the time when infants are capable of distinguishing strangers from familiar persons and the time when they begin to show fear responses.

Infants' fear of strangers is widespread and occurs cross-culturally (Smith, 1979). Although this is an intensively studied developmental phenomenon, almost no attention has been paid to its biological underpinnings. Nor has much attention been paid to characteristics of strangers that elicit the most intense response. We predict that infant fears should be more intense for strange males than for strange females because of the greater potential for harm associated with aggression by unknown and unrelated males.

Numerous studies support this prediction. Young children are likely to experience more harm from males than from females and more harm from unrelated males than from related males (Daly & Wilson, 1988). Although we do not know whether such violence was common in our evolutionary past, studies of primates show that male aggression toward infants is widespread (Hrdy, 1981). Furthermore, violent behavior is much more likely to come from strange males when they take over a troop. Marks (1987, p. 23) suggests that stranger fear is "an evolutionary remnant reflecting widespread abuse and infanticide by strangers during the evolution of hominids and their predecessors."

Studies that have assessed the gender of the stranger have found that males are more likely to be feared than females (Smith & Sloboda, 1986; Morgan & Ricciuti, 1967; Greenberg, Hillman & Grice, 1973; Solomon & DeCarie, 1976; Skarin, 1977). The study by Smith and Sloboda (1986) found a mean negative response to all male strangers, compared with a mean positive response to female strangers. The results of research on gender-related responses to strangers cannot be explained by height differences between males and females because tall females do not elicit the same degree of fear as males do (Horner, 1981). Also, it cannot be explained by differences in facial hair because infants whose fathers had beards were also afraid of strange males with beards (Horner, 1981).

Researchers have explained the differential response to males as being due either to a general tendency of females to be friendlier toward and more interested in infants or by a "discrepancy hypothesis." That is, strange males are more different from mothers than are strange females, and it is this discrepancy to which infants are responding. However, if discrepancy were the cause, then infants should also fear older children, who are also different. This is not the case. In fact, infants often respond positively to other children (Brooks & Lewis, 1976; Horner, 1981; Lewis & Brooks, 1974). In one study, seven- to 19-month-old infants who responded negatively to strange adults were neutral or slightly positive when approached by a strange child (Greenberg et al., 1973). Smith reports that one-year-olds show an interest in age peers and seek "tentative affiliation," often mediated by a toy (Smith, 1979). Garvey (1990) notes that toddlers often sit next to each other but interact more with objects than with each other. Lewis and Brooks (1974) also found that year-old infants show a special interest in others their age.

Although we did not make any predictions about infants' response to other children, their attraction to other children may also serve as a safety function. By seeking to affiliate with other children, an infant who has been separated from its mother would be less likely to be harmed than when it is alone, as predicted by the "selfish herd" theory (Hamilton, 1971). Groups have more eyes and ears and also reduce the likelihood of attack to 1/N, where N is the size of the group.

Avoiding Dangerous Animals Animals were likely to have been important sources of danger in ancestral environments. Hazards included small animals (poisonous snakes and spiders), carnivores (such as wild dogs, hyenas, and large cats), and other large animals. Thus, strong selective pressure for the development of predator-detection and predator-avoidance mechanisms has existed for a long time. However, specific fears should differ with age because risks and vulnerability are age-dependent.

Therefore, we predict that the onset of animal fears should be associated with independent movement, which increases the likelihood of coming into contact with animals.

Young children, when they are beginning to crawl and when they first walk, seldom venture very far from home. Within this space they are more likely to come into contact with small animals such as insects, spiders, and snakes than with large predators. Large animals, especially predators, are less numerous and less likely to be near to camp where they could be encountered by young children. Therefore, we predict that children should show fear responses associated with small dangerous animals before they show fear of large animals. Specifically, we predict that fear of spiders, snakes, and other small animals should begin before three years of age and be apparent when children are still crawling or just beginning to walk.

Studies of fears of children younger than three years old have focused more on responses to strangers and social situations than to animals. Research on infant and toddler fear of animals is drawn largely from interviews with their mothers rather than from direct observations. There is some experimental evidence, however, that suggests that infants fear conditions and stimuli that are likely to be associated with dangerous small animals, such as approaching objects and sudden movement, especially movement toward the child (Scarr & Salapatek, 1970; Smith, 1979; Russell, 1979). Sudden erratic movement is a characteristic of animals that have been flushed from the cover of vegetation. These studies also show that fear responses to these stimuli begin to appear at about seven months. This supports our prediction that fears of small animals would be manifest with the onset of crawling.

Much research on children's fears, unfortunately, does not regularly distinguish the kinds of animals that children most fear. Research articles tend to combine all animal fears. However, there is some support for the earlier development of fear of spiders and snakes than fear of large animals. For instance, Agras, Sylvester, and Oliveau (1969) found that fear of snakes begins around age two and continues to rise until about age 12. Thereafter it slowly declines, but it continues to remain high even into adulthood. Muris, Merckelbach, and Collaris (1997) found that the fear of spiders begins around the age of 3.5 and continues to be high throughout childhood. Lapouse and Monk (1959) also found that fear of "bugs" was greater among children six to eight than among those ages nine to 12.

Studies that distinguish among types of animals show that fear of large animals or predators is common among children more than four years old (Ollendick et al., 1985; Muris et al., 1997; Angelino, Dollins & Mech, 1956; King, Hamilton & Ollendick, 1997). Specific animals mentioned include dogs, bears, and wolves. Muris et al. found that fear of "predators" was common in children ages nine to 13. Contrary to our predictions, however, fear of dogs has been found in children as young as two (Hagman, 1932; Holmes, 1935) as well as in older children. Large animal fears begin to wane after about 10 years of age, when social fears become much more prominent (Angelino et al., 1956; Croake, 1969; Scherer & Kakamura, 1968; King et al., 1988; Ollendick et al., 1985).

Although we had not predicted this, the greatest fears of children less than five years old were of darkness, the supernatural (ghosts and witches), storms, and being alone (Jersild & Holmes, 1935; Holmes, 1935; Jersild, Markey & Jersild, 1933; Hagman, 1932; Bauer, 1976). Even children older than six are afraid of woods and dark places, derelict places, and unknown distant places (Hart, 1979). Such conditions are likely to be associated with hazards. For instance, ghosts and witches in children's stories are almost always connected with darkness, deep woods, and being alone. These conditions are often associated with dangerous animals, sudden attack, or other threatening events (Russell, 1979). Russell argues that predator-avoidance mechanisms produced by spatial, visual, and sensory conditions reliably associated with dangerous animals are more likely to evolve than fears of specific animals. Given that the kinds of animals likely to be encountered differ geographically, we concur.

Avoiding Topographic and Natural Hazards Natural hazards are relevant across the childhood age span. Hazards that have been a significant force throughout our evolutionary history include fire, deep and fast-moving water, heights with sudden drop-offs (such as cliff and mountain edges), storms (especially intense winds, thunder, and lightening), and fire. Weather, earthquake, and fire hazards often come suddenly and necessitate fast response, such as finding shelter, getting help, or moving away from the hazard. Regardless of their age, children would be at risk if they did not respond appropriately to these hazards.

Therefore, we predict that fears of such phenomena as storms, earthquakes, fire, and water will begin early, remain high throughout early and middle childhood, and wane with the teen years when children are more capable of anticipating dangers and taking actions to protect themselves.

Studies of children's fears show that infants respond fearfully to stimuli associated with storms, such as loud noises and bright lights (Scarr & Salapetek, 1970). However, we are aware of no studies of infants' responses to other natural hazards or stimuli that are typically associated with hazardous conditions.

Studies of older children show that storms and fire are among the most common fears of preschoolers (Jersild & Holmes, 1935; Hagman, 1932). Others have found that children ages six to 12 fear earthquakes, fire, thunder and lightning, and deep water and drowning (Marks, 1987; Lapouse & Monk, 1959; Scherer & Kakamura, 1968; Ollendick, Matson & Heisel, 1984; King et al., 1988). Topographic hazards such as fears of heights and falling do not appear to be as common as fears of storms, fire, water, and earthquakes (Ollendick et al., 1984). Many studies have found discrepancies between child and parental fears. For instance, Moore and Young (1978) found that parents most feared traffic and attack by unfriendly humans. Environmental fears such as large animals, snakes, spiders, open water, and high places were much lower on parents' lists, even though they were highly salient to children.

Results of these studies are consistent with our prediction, but they do not constitute strong support because our prediction is very general. However, if striking age-related onsets or terminations of fears of physical hazards exist, they would constitute strong evidence against it.

Finding and Constructing Shelter As children begin to move independently and explore the environment, they should be motivated to seek out spaces that afford safety and protection. According to Appleton (1975), an effective refuge is one that offers protection overhead but has permeable boundaries that enable viewing out. An example of a natural refuge would be a tree with a spreading canopy or a shrub that is open enough to allow a child to sit within it. Given the highly focused nature of many children's play activities, being in a safe place during play would leave them less vulnerable to predation than would playing in open, unprotected areas. Shelter is especially important for play when children's attention is focused on their activities and not on the surrounding environment. Thus, children are more likely to fail to detect predators or hostile people when playing intently with an object or with another child than when they are actively moving and attending to the environment.

Therefore, we predict that young children, especially of preschool age, will seek out naturally occurring shelter in the environment and that older children will actively shape or construct shelters.

Evidence to test these predictions comes from studies of children's naturally occurring play behaviors, both outdoors and indoors. Research on preschool settings shows that small-scale semienclosed areas that accommodate two to five children are the most popular zones in playrooms and are associated with more affiliative behaviors, cooperative play, and sustained activity than spaces that lack these features (Legendre & Fontaine, 1991; Moore, 1986; De Long, 1991). Lowry (1993), whose research focused on the value of solitary play and stimulus retreat, found that preschool-age children were more likely to engage in focused, solitary play in small play structures that had greater enclosure (such as a roof and sides with a view out versus a roof only) than in open rooms lacking small-scale enclosure.

These results have stimulated design recommendations for daycare centers that call for partial acoustic and visual dividers between spaces that create small, semienclosed areas (Cohen, McGinty, Armstrong & Moore, 1982). Olds (1987) specifically recommends boundaries that are permeable and transparent enough to allow children to be safe and to be able to watch caretakers and other children. In fact, if a space becomes too enclosing and if visual obstacles reduce children's ability to see caretakers, they are less likely to use the space (Legendre & Fontaine, 1991). Appleton (1975) would argue that such spaces fail as refuges because they reduce the ability of individuals in them to survey the surrounding environment from a position of safety.

When playing outdoors, children show a strong attraction to refuge-type settings during the early childhood years (Kirkby, 1989; Hart, 1979; Matthews, 1992). Kirkby found that children were attracted to a "scruffy" part of a schoolyard, lined with overgrown junipers and shrubbery. Over the years, so many children had played in the area that there were pathways, tunnels, and small "rooms" throughout the vegetation. Imagine the children's disappointment when they found the vegetation totally cleared away one Monday morning, the result of an adult landscape "clean-up" party. Grass was planted in place of the shrubs.

Kirkby's research is significant because she considered the nature of children's play. She predicted that children would engage in more imaginative and dramatic play in a natural refuge because it afforded a greater sense of enclosure and more opportunities to manipulate objects than a built refuge on the playground. Her analysis shows that dramatic play ranged from 42 percent of the total play content in the built refuge to

68 percent in the natural refuge settings. A detailed spatial analysis of the natural refuge settings showed that they contained a number of small subspaces appropriate for two to three children, with varying degrees of visual openings and lookouts. The built structure, in contrast, was much more visually open and less complex, and its components could not be manipulated. Kirkby (1989, p. 11) writes: "Some of the finest examples of children's play occurred in the smallest spaces, where the children seemed less distracted and more engrossed. The enclosure itself, by cutting off peripheral stimuli, enhanced their ability to engage in dramatic play." In a natural history of children's outdoor play, Hart (1979) found that preschool-age children preferred to play in the dirt underneath the trees and in small, enclosed spaces. Large boxes in particular were favorite play items, and children often turned these into playhouses and forts.

Our prediction that older children would prefer to build shelters is supported by existing data (Hart, 1979; Miller, 1984). Hart found that preferred outdoor activities for children between the ages of seven and 11 included building forts and tree houses. He found that children spend large amounts of time modifying the landscape to make places for themselves and their play. The youngest children were more likely to make subtle modifications using dirt, grass, or weeds. The older children, on the other hand, built things or actively transformed vegetation to make burrowlike forts. Both older and younger children used bushes for shelter, particularly large bushes with lush canopies and an open network of branches. The fort-building activities usually required several children working together, whereas the younger children often used existing places that they find on their own. Boys did almost all of the constructing. Girls tended to make small things used in decorating the forts and shelters.

Miller (1984) found similar results in her study of children's use of a wooded area outside their school. The children, ages seven through nine, spent recesses and after-school hours in the woods, building and modifying a series of forts and secret places that had obviously been passed down through the years. Some of these were nestlike modifications to vegetation, whereas others were more elaborate constructions built from branches and other natural materials. The woods were used for play activities much more heavily than the playground area at the school. Miller also found that children spent a great deal of time making new paths and trails that provided multiple ways of moving through the environment and multiple viewing locations for "spy" games. These data are consistent with research on wayfinding and visual surveillance (Orians & Heerwagen, 1992).

Predictions Related to Foraging and Feeding

Major ecological tasks for all animal species are to find food and to distinguish edible from inedible items. Generalized omnivorous species eat a large number of food types, and what is available varies greatly geographically. Therefore, members of such species rarely have innate food-recognition abilities. Humans do not possess innate food-recognition mechanisms. Rather, they learn what is edible by observing, by sampling items in their environments, and by being instructed. However, sampling is risky because toxic materials may be ingested. Thus, we would expect mechanisms to evolve that reduce the potential for ingesting harmful items while still enabling exploration and learning.

The following factors are likely to influence the timing and nature of children's ability to identify edible foods and to distinguish them from items that are harmful. First, children need to know which objects are edible by the time weaning occurs because at that point they will no longer have breast milk available as a primary source of nutrition. Second, because breast milk has antiobiotic properties, nursing infants can sample unfamiliar objects with less risk than children who have been weaned. The third factor affecting the timing and nature of food identification is children's mobility. A relatively safe way for children to learn what is edible is to try out a variety of substances within view of caretakers. Children are likely to be in the presence of caretakers when their mobility is low. When children are able to walk and move rapidly out of view, they are more likely to sample foods where adults cannot see and advise them.

Therefore, we predict that infants will begin to put strange objects in their mouths as soon as they can move about and that they will cease doing so at about the time they are weaned and begin to regularly consume foods other than breast milk.

Studies of object exploration show that mouthing behaviors begin at about two months of age, prior to crawling. At this age, objects can be used to pacify crying infants or to provide visual interest (Uzgiris, 1967). Mouthing increases up until about six months, then begins to decrease, and is rarely observed after two years (Uzgiris, 1967). Although research on this topic is not extensive, the available data do support our predictions.

Even though mouthing small objects is a well-documented infant behavior, there is very little explanation for why it occurs. In addition to the predictions we make here, other authors have suggested that mouthing of small objects from the surrounding habitat enables young children to ingest helpful microbes to replace those in breast milk when they are weaned (Neel, 1970).

Predictions Related to Finding a Place to Live

Our ancestors lived in environments devoid of modern comforts and conveniences. Their survival, health, and reproductive success depended on their ability to understand relationships between habitats and resources and to evaluate habitat quality. However, this knowledge unfolds gradually as children experience the environment at ever increasing scales. Children less than two years of age seldom wander more than a couple of hundred feet from their mothers when they are playing (Anderson, 1972). The range increases steadily over the next 10 years as children develop physical skills for long-distance movement and cognitive skills that help them orient in the environment and find their way home (Matthews, 1992; Hart, 1979; Moore & Young, 1978). The play range of children ages three to five is limited to the home and the front yard. By the age of six to nine, however, children's range is up to 10 times greater, and they actively explore places beyond their home, although they are generally within calling range of a parent. By ages 10 to 12, both the territorial range and number of destinations increase dramatically. Studies of contemporary hunter-gatherer societies show similar expansion of children's range from infancy to adolescence (Konner, 1976; Draper, 1976).

Distant objects and landscape patterns would not have been relevant to young children whose activities were confined to their camp and its immediate surroundings. During seasonal shifts of campsites, young children would have been in the company of adults and would have had little need to observe where they were going or how to select a travel route. As we have already shown, younger children selectively pay attention to nearby, small objects even though they are capable of perceiving distant objects. However, when children begin to engage in behaviors that remove them from their social group and during which they need to navigate over a broader terrain, paying attention to distant objects and spatial relationships among objects acquires value.

Therefore, we predict that interest in landscapes will develop quickly at about the age of adolescence. From an evolutionary perspective, this is the age at which young men would have begun to hunt and at which young women would have been likely to leave their natal group and begin families. We further predict that adolescents, both males and females, should prefer savanna landscapes over other types of habitats. Because a relatively small number of generations has transpired since *Homo sapiens* occupied temperate habitats, it is reasonable to postulate that landscape features characteristic of high-quality African savannas, in which our species evolved, would be especially attractive to humans today (Orians, 1980).

Unfortunately, little research on the onset of attention to landscapes has been conducted. Most research on responses to landscapes has focused on adults or college-level teenagers; little has been done on children between the ages of six and 13.

None of the few studies available was conducted to specifically test our prediction, but several report relevant results. Balling and Falk (1982) compared responses of children to a variety of vegetation types, some familiar and some not. Their age groups consisted of students in grades three, six, and nine, college students, adults, and retired citizens, all Americans. The third-graders (eight-year-olds) gave the highest preference scores to all landscapes, including the savanna, and the ninth-graders gave the lowest scores. This pattern is contrary to our prediction that preferences for landscapes, especially the savanna, would be evident at adolescence.

Lyons (1983) also compared responses to slides of landscapes by children ranging from grades three (mean age 8.5 years) through nine (mean age 14.6 years) with responses of college-age students and adults. She also compared responses of males and females. Third-graders consistently gave scenes of landscapes the highest scores; mean scores declined with age from grades three through nine. Scores stabilized for college-age students and young adults but declined to their lowest level for elderly (mean age 59.5 years) adults. She found no significant differences between the sexes in responses to landscapes. Similar age-related results were reported in a study of Herzog, Herbert, Kaplan, and Crooks (2000) that compared responses of Americans and Australians to vegetation types.

In these three studies, teenagers gave lower preference scores to landscapes than younger children did, the opposite of our prediction. Therefore, our prediction may be incorrect, but further research is needed before definitive conclusions are reached. The meaning of the responses of children to questions about their preferences for varied landscapes is unclear and open to multiple interpretations. For example, younger children may be more inclined than teenagers to respond in ways they think will please the experimenters. Also, teenagers may be more responsive to the social environment than to the physical environment and may find landscapes less inviting than highly social places. More imaginative ways of evoking responses to landscapes among young people need to be devised before we have a clear idea of how children react to and express interest in the scenery around them.

Implications of the Evolutionary Approach to Modern Life

The research we have reviewed on children's fears and preferences raises several issues relevant to modern life styles. In this section we first discuss the potential disconnection between children's fears related to hazards in our evolutionary past and modern-day dangers. Second, we consider how an ecological-evolutionary approach could be used to design environments that are more appealing and better matched to the developmental stages of children. Third, we look at how the electronic media, particularly video games, capitalize on children's intrinsic fears and attractions.

Some of the childhood behaviors we have described are no longer adaptive in modern society and may even be potentially harmful. For example, mouthing of objects is dangerous today because many poisonous materials are present in the environment in and around homes in modern societies. During the hunter-gatherer era, most of these substances did not exist, and some protection was provided by the antibiotic properties of breast milk. Furthermore, the low population densities and regular movements of people reduced the presence of disease organisms.

Research also shows that fears are more strongly associated with hazards of our evolutionary past rather than with real dangers today. Children are more likely to be harmed by modern objects such as guns and cars than by snakes, spiders, and large predators. Nonetheless, children continue to manifest strong fears of these biological dangers as well as fears of darkness, monsters, and being alone, even in their own homes with their parents nearby. These fears are powerful enough to warrant psychological intervention in many instances. The fears, of course, are likely to be influenced by cultural reinforcement and generalized expectancies (Davey, 1995), as well as by the evolutionary pressures we discuss. However, research on phobias has shown that one does not need to have experience with an object to develop a strong fear of it (Muris, Merckelbach & Collaris, 1997; Poulton et al., 1998). For example, laboratory studies of fear acquisition and extinction show that it is much easier to acquire fear responses and more difficult to extinguish them when stimuli are evolutionarily relevant (snakes or spiders) as compared to neutral stimuli or to modern hazards such as guns (Cook et al., 1986; Hugdahl & Karker, 1981). Although these studies were conducted with adult subjects, there is no reason to believe results would be different with children.

Even more striking, from an evolutionary point of view, are the results of "backmasking" experiments in which slides are displayed subliminally (15 to 30 milliseconds) before being "masked" by a slide of another stimulus or setting. Even though the subjects are not consciously aware of having seen the stimulus slide, presentations of natural settings that contain snakes or spiders elicit strong aversive or defensive reactions in nonphobic persons (Öhman, 1986; Öhman & Soares, 1993). In persons who already have phobias, a masked subliminal presentation is sufficient to elicit defensive responses to the feared stimulus (Öhman & Soares, 1994). Thus, experimental results show that aversive responses can occur without recognition or awareness of specific natural threat stimuli but that no such responses exist to neutral stimuli or to modern hazards. These results are inconsistent with a purely learned or cultural interpretation of phobias and fears.

Although we have identified a number of studies on children's preferences and fears related to the natural environment, very little of this research is used in the design of outdoor playgrounds or indoor spaces. An important exception is the work by Cohen and her colleagues (Cohen et al., 1987) and by Moore and Young (1978). The design of daycare centers, playgrounds, schools, homes, and hospitals could benefit enormously from a better understanding of children's natural play behaviors. Even a cursory investigation of schools and playgrounds shows that little has changed over the past 50 years. Children still sit in desks facing a teacher or sometimes in clusters of desks. And they still play in environments dominated by swings and slides or other fixed play equipment that does little to capture their imagination.

As we noted in the discussion on young children's attraction to natural refuges in playgrounds, play behaviors in natural refuges differed significantly from play behaviors in built refuge or traditional playground equipment. The natural refuges and natural materials (flowers, sticks, stones) facilitated long bouts of imaginary play, a behavior with high social and cognitive payoffs (Kirkby, 1989; Moore, 1989; Hart, 1979). We also discussed several studies of children in daycare settings. These studies showed that refuge settings, with views to primary caretakers, seem to give children a sense of security that facilitates extended play and exploration with toys (Cohen et al., 1987; Olds, 1987). The integration of research into design has been most successful with young children in preschool environments. In contrast, the design of elementary or secondary school environments has paid little heed to research on children's play. Even though there is good evidence of the value of natural landscapes and moveable objects for constructing, schools still have playgrounds with hard surfaces and no vegetation. To some extent, this is due to fears of litigation associated with potential accidents from construction activities. But educators are also concerned about maintenance and surveillance. As noted in the research by Kirkby (1989), the natural refuge areas of the playground she studied were removed because they were difficult to maintain and also because the teachers could not see into the refuges. Unfortunately, the sense of privacy and enclosure contributes to the appeal of refuges from the children's perspective.

In contrast to the situation in the physical environment, toy and video game manufacturers have invested a great deal of time and money in learning more about the features of the virtual environment that capture and sustain attention. Video games feature places, stimuli, and events with strong roots in our evolutionary past—predators, prey, ominous strangers, natural hazards, scary places, ghosts, poisonous foods, and a safe home base. Researchers have found that the most appealing games have a cluster of motivating features, including dynamic visual imagery, randomness, action, dynamic hazards, spatial and visual complexity, audio effects, fantasy, constant feedback, and interactivity (Greenfield, 1984; Loftus & Loftus, 1984; Griffiths, 1996; Sheldon, 1998). According to Malone (1981), the goals, challenge, and fantasy of video games feed intrinsic motivation, thereby increasing the desire to continue playing. No wonder these games are so compelling in contrast to ordinary environments.

Given the powerful appeal of these games, many educators and parents are concerned about the impact of video games on reading skills and social aggression from modeling or desensitization (Greenfield & Cocking, 1994). However, results of studies in this area are mixed. Some find increased aggression in children after playing games with high violent content, while others do not (Griffiths, 1996). One of the more curious aspects of video games is the relative lack of interest and lower game skills shown by girls, compared to boys. Numerous studies show that girls respond negatively to the violent game themes (Greenfield, Brannon & Lohr, 1994). However, even when the violent content is controlled, girls still do not perform as well on the games as do boys (Greenfield, 1994).

Although most research has focused on the potential downside of video games, there is also increasing interest in identifying benefits of game playing. Laboratory studies show that video games can improve spatial visualization and mental rotation skills (Okagaki & Frensch, 1994) as well as attentional focus and the development of logical and strategic planning skills (Blumberg, 1998; Mandinach & Corno, 1985), skills that are increasingly important to all sorts of computer applications in a technological world (Greenfield, 1994). Greenfield (1994, p. 5) sees video games as "revolutionary in that they socialize children to interact with artificial intelligence on a mass scale and from a very early point in their development." If this is so, it represents a fundamental departure from our evolved, face-to-face interactions with people, places, and other species that have characterized our relationships with the environment throughout human history. It is clear from the evidence on video games that children are drawing on skills that have been useful in exploring the environment and staying safe and yet that these skills and motivations are becoming increasingly detached from physical and social reality.

Even as evidence of enhanced cognitive skills associated with video games is beginning to appear, there is also evidence that children feel increasingly isolated and use games and TV as "electronic friends" (Griffiths, 1996). Research on children's television viewing habits shows that children are attracted to TV because it provides companionship and the potential to learn about human relationships and to garner information that is useful in solving personal problems (Sheldon, 1998). In a sense, TV has become for our children what the campfire might have been for our ancestors. It is a way to relax at the end of the day and find out what is going on. As Sheldon notes, the characters in children's favorite TV shows form the basis for shared interests and topics of conversation with their peers at school the next day.

Studies of how young children spend their time also show that a large part of the day is spent watching TV or playing video games (Huston, Wright, Marquis & Green, 1999) rather than engaging in imaginative or constructive play, especially in outdoor settings. TV displaces outdoor and social activities (Williams & Handford, 1986). Australian children in a town without TV spent more time playing and doing outdoor activities than in similar towns with TV (Murray & Kippax, 1978). Experiments in reduced TV time also show that children increase their creative play activities and reading (Gadberry, 1980).

Outdoor play, as many researchers have pointed out, may be especially valuable because it integrates cognitive, emotional, and social behaviors (Hart, 1979; Moore & Young, 1978; Stutz, 1996). For example, construction and imaginative play require planning skills, coordination, negotiation, creativity, and trust, which form the basis for teamwork. Often these activities involve children of different ages and thus provide learning and socialization opportunities that may be more compatible with our evolutionary heritage than today's planned outdoor activities that limit interaction to same-age children. As noted by Draper (1976) in her studies of the !Kung San, children in hunter-gatherer groups played in mixed-age groups regularly because there were relatively few children in each group. Outdoor play also is more likely to be compatible with children's developmental stage, unlike TV programs that contain themes, images, and concepts more suitable for mature audiences (Stutz, 1996). In addition, outdoor play often has a slower, less programmed rhythm associated with exploration and discovery. The psychological value of aimless exploration, especially in natural settings, may be more important than many realize. It is associated with creativity (Claxton, 1997), stress reduction (Ulrich, 1993), and self-esteem (White & Heerwagen, 1998).

In addition to the developmental issues of natural play, it is also worth asking what impact children's increasing distance from the natural world will have on their attitudes and feelings about the environment. Autobiographies of many of the leaders in the environmental field show a strong and early attachment to the natural world (Wilson, 1994). Can today's children really learn to value nature when they are so removed from the natural world? A study by Hoyt and Acredolo (1992) shows that environmental attitudes, preferences for nature, and the development of pastoral values are strongly influenced by children's actual experience in natural settings. In contrast, urbanism was positively related to preferences for built settings and negatively related to preferences for natural environments. In addition to increased positive attitudes about nature, knowledge about biodiversity also appears to be influenced by experience of nature. Chipeniuk (1995) found that children's foraging behavior in natural environments was related to increased knowledge of biodiversity of habitats in adult life. As they forage, children learn a great deal about the distribution of plants, the numbers of species growing in any given location, and the relationship between different species.

There is a growing concern among psychologists that video games may contribute to psychological distancing and social aggression (Alloway & Gilbert, 1998; Newson, 1996). Can this kind of psychological and emotional distancing also occur in relationship to the environment? As children's play becomes increasingly virtual, we need to know more about the consequences for the development of environmental knowledge and values and, ultimately, the willingness to protect the natural world. A recent study by Levi and Kocher (1999) lends credence to this concern. Levi and Kocher assessed the impact of experiencing nature through virtual reality. They found that simulated experience with spectacular natural environments, as depicted through commercially available products, increased support for national parks and forests but decreased support for local natural areas. The devaluing of nonspectacular, local environments is disturbing because these landscapes play a key role in global ecology. Levi and Kocher also found that simulated experience with nature had positive psychological benefits to the individuals viewing the scenes, a result that has been found by many other researchers (see Ulrich, 1993, for a review). As virtual reality becomes more commercially viable and thus available to a wider audience, the costs and benefits of simulated experience will become more important. On the one hand, virtual nature can be used therapeutically to reduce stress and provide contact with natural environments that people are not likely to be able to experience on their own—especially those who are ill or elderly. On the other hand, by increasing experience of nature at a distance, support for preservation of nonspectacular but ecologically valuable landscapes may diminish.

Conclusion

Although this chapter contains no original research data, we believe that we have presented enough evidence to demonstrate the explanatory power of an evolutionary approach to the development of childhood behavior and to show its ability to help guide a comprehensive research agenda. Children clearly have adapted behaviorally to their ecological worlds, and they show evidence of those adaptations in modern life, both in their play behaviors and in their fears. We need to better understand the details of these adaptive responses and their implications for children's lives in an increasingly technological world dominated by electronic relationships and virtual environments.

Acknowledgments

We wish to thank Peter H. Kahn, Jr. and Stephen R. Kellert for their insights, support, and guidance throughout the development of this chapter. We also wish to thank Roger Ulrich for many stimulating discussions of these issues over the past several years.

References

Agras, S., Sylvester, D., & Oliveau D. (1969). The epidemiology of common fears and phobia. *Comprehensive Psychiatry, 10*, 151–156.

Alloway, N., & Gilbert, P. (1998). Video game culture: Playing with masculinity, violence, and pleasure. In S. Howard (Ed.), *Wired up: Young people and the electronic media*. London: UCL Press.

Anderson, J. W. (1972). Attachment behaviors out of doors. In N. Blurton Jones (Ed.), *Ethological studies of child behavior*. Cambridge: Cambridge University Press.

Angelino, H., Dollins, J., & Mech, E. V. (1956). Trends in the "fears and worries" of school children as related to socio-economic status and age. *Journal of Genetic Psychology, 89*, 263–276.

Appleton, J. (1975). *The experience of landscape*. New York: Wiley.

Archer, J. (1979). Behavioral aspects of fear. In W. Sluckin (Ed.), *Fears in animals and man*. New York: Van Nostrand Reinhold.

Balling, J., & Falk, J. (1982). Development of visual preference for natural landscapes. *Environment and Behavior, 14*, 5–28.

Bauer, D. H. (1976). An exploratory study of development changes in children's fears. *Journal of Child Psychiatry, 17*, 69–74.

Berk, L. (1997). *Child development* (4th ed.). Boston: Allyn and Bacon.

Blumberg, F. C. (1998). Developmental differences at play: Children's selective attention and performance in video games. *Journal of Applied Developmental Psychology, 19*, 615–624.

Bogin, B. (1991). *Patterns of human growth* (2nd ed.). Cambridge: Cambridge University Press.

Bower, T. G. R. (1974). *Development in infancy.* San Francisco: Freeman.

Bowlby, J. (1980). *Attachment and loss.* New York: Basic Books.

Bronson, G. W. (1972). Infants' reactions to unfamiliar persons and novel objects. *Monographs of the Society for Research in Child Development, 37*(3).

Brooks, J., & Lewis, M. (1976). Infants' responses to strangers: Midget, adult and child. *Child Development, 47*, 323–332.

Chase, R. A. (1992). Toys and infant development: Biological, psychological and social factors. *Children's Environments, 9*, 3–12.

Chipeniuk, R. (1995). Childhood foraging as a means of acquiring competent human cognition about biodiversity. *Environment and Behavior, 27*, 490–512.

Chomsky, N. (1975). *Reflections on language.* New York: Pantheon.

Clarke-Stewart, K. A. (1973). Interactions between mothers and their young children: Characteristics and consequences. *Monographs of the Society for Research in Child Development, 38*(153).

Claxton, G. (1997). *Hare brain/tortoise mind: How intelligence increases when you think less.* New York: Harper Collins.

Cohen, U., McGinty, T., Armstrong, B. T., & Moore, G. T. (1987, Fall). The spatial organization of an early childhood development center: Modified open space, zoning and circulation. *Daycare Journal*, 35–38.

Cook, E. W., Hodes, R. L., & Lang, P. J. (1986). Preparedness and phobia: Effects of stimulus content on human visceral conditioning. *Journal of Abnormal Psychology, 95*, 195–207.

Cosmides, L., & Tooby, J. (1993). The psychological foundations of culture. In J. Barkow, J. L. Cosmides & J. Tooby (Eds.), *The adapted mind: Evolutionary psychology and the generation of culture.* New York: Oxford University Press.

Croake, J. W. (1969). Fears of children. *Human Development, 12*, 239–274.

Daly, M., & Wilson, M. (1988). Evolutionary social psychology and family homicide. *Science, 242*, 519–524.

Davey, G. C. (1995). Preparedness and phobias: Specific evolved associations or a generalized expectancy bias? *Behavioral and Brain Sciences, 18*, 289–325.

De Long, A. J. (1991). Enhancing learning in child care centers through design. In A. F. Torrice & R. Logrippo (Eds.), *Design of the times: Day care.* Burlingame, CA: Living and Learning Environments.

Draper, P. (1976). Social and economic constraints on child life among the !Kung. In R. B. Lee & I. DeVore (Eds.), *Kahlahari hunter-gatherers: Studies of the !Kung San and their neighbors.* Cambridge, MA: Harvard University Press.

Feiring, C., Lewis, M., & Star, M. D. (1984). Indirect effects and infants' reactions to strangers. *Developmental Psychology, 20*, 207–212.

Gadberrry, S. (1980). Effects of restricting first graders' TV viewing on leisure time use. *Journal of Applied Developmental Psychology, 1*, 45–58.

Garvey, C. (1990). *Children's play*. Cambridge, MA: Harvard University Press.

Greenberg, D., Hillman, J. D., & Grice, D. (1973). Infant and stranger variables related to stranger anxiety in the first year of life. *Developmental Psychology, 9*, 207–212.

Greenfield, P. M. (1984). *Mind and media: The effects of television, video games and computers*. Cambridge, MA: Harvard University Press.

Greenfield, P. M. (1994). Video games as cultural artifacts. *Journal of Applied Developmental Psychology, 15*, 3–12.

Greenfield, P. M., Brannon, C., & Lohr, D. (1994). Two-dimensional representation of movement through three-dimensional space: The role of video game expertise. *Journal of Applied Developmental Psychology, 15*, 87–103.

Greenfield, P. M., & Cocking, R. R. (1994). Effects of interactive entertainment technologies on development. *Journal of Applied Developmental Psychology, 15*, 1–2.

Griffiths, M. (1996). Computer game playing in children and adolescents: A review of the literature. In T. Gill (Ed.), *Electronic children: How children are responding to the information revolution*. London: National Children's Bureau.

Hagman, E. R. (1932). A study of fears of children of pre-school age. *Journal of Experimental Education, 1*, 110–130.

Hamilton, W. D. (1971). Geometry for the selfish herd. *Journal of Theoretical Biology, 31*, 295–311.

Hart, R. (1979). *Children's experience of place*. New York: Irvington.

Heft, H. (1988). Affordances of children's environments: A functional approach to environmental description. *Children's Environments Quarterly, 5*, 29–37.

Herzog, T. R., Herbert, E. J., Kaplan, R., & Crooks, C. L. (2000). Cultural and developmental comparisons of landscape perceptions and preferences. *Environment and Behavior, 32*, 323–346.

Hill, K., & Hurtado, A. M. (1996). *Ache life history: The ecology and demography of a foraging people*. New York: Aldine de Gruyter.

Holmes, F. B. (1935). An experimental study of the fears of young children. *Child Development Monographs, 20*, 167–296.

Horner, T. M. (1981). Two methods of studying stranger reactivity in infants: A review. In S. Chess & A. Thomas (Eds.), *Annual progress in child psychiatry and child development*. New York: Bruner/Mazel.

Hoyt, K. A., & Acredolo, L. P. (1992). How to Childhood Experiences Influence Environmental Attitude Formation? In *Proceedings of the Environmental Design Research Association Confluence, 23* (pp. 221–228). Bouldn, CO, April 9–12.

Hugdahl, K. (1978). Electrodermal conditioning to potentially phobic stimuli: Effects of instructed extinction. *Behavioral Research and Therapy, 16*, 315–321.

Hugdahl, K., & Karker, A. C. (1981). Biological vs. experiential factors in phobic conditioning. *Behavioral Research and Therapy, 19*, 109–115.

Huston, A. C., Wright, J. C., Marquis, J., & Green, S. B. (1999). How young children spend their time: Television and other activities. *Developmental Psychology, 35*, 912–925.

Hrdy, S. (1981). *The woman who never evolved*. Cambridge, MA: Harvard University Press.

Jersild, A. T., & Holmes, F. B. (1935). Children's fears. *Child Development Monographs, 20*. New York: Teachers College, Columbia University.

Jersild, A. T., Markey, F. U., & Jersild, C. L. (1933). Children's fears, dreams, wishes, daydreams, likes, dislikes, pleasant and unpleasant memories. *Child Development Monographs, 12*. New York: Teachers College, Columbia University.

Katz, D. R. (1990, February). The new generation gap. *Esquire*, 49–50.

Kendler, K. S., Neale, M. C., Kessler, R. C., Heath, A. C., & Eaves, L. J. (1992). The genetic epidemiology of phobias in women. *Archives of General Psychiatry, 49*, 273–281.

King, N. J., Hamilton, D. I., & Ollendick, T. H. (1988). *Children's phobias: A behavioral perspective*. New York: Wiley.

King, N. J., Ollendick, T. H., & Hurphy, G. C. (1997). Assessment of childhood phobias. *Clinical Psychology Review, 17*, 667–687.

Kirkby, M. (1989). Nature as refuge in children's environments. *Children's Environments Quarterly, 6*, 1–12.

Konner, M. (1976). Maternal care, infant behavior and development among the !Kung. In R. B. Lee & DeVore, I. (Eds.), *Kahlahari hunter-gatherers: Studies of the !Kung San and their neighbors*. Cambridge, MA: Harvard University Press.

Lapouse, R., & Monk, M. A. (1959). Fears and worries in a representative sample of children. *American Journal of Orthopsychiatry, 29*, 803–818.

Lee, R. B., & DeVore, I. (Eds.). (1976). *Kahlahari hunter-gatherers: Studies of the !Kung San and their neighbors*. Cambridge, MA: Harvard University Press.

Legendre, A., & Fontaine, A. M. (1991). The effects of visual boundaries in two-year-olds' playrooms. *Children's Environments Quarterly, 8*, 1–16.

Levi, D., & Kocher, S. (1999). Virtual nature: The future effects of information technology on our relationship to nature. *Environment and Behavior, 31*, 203–226.

Lewis, M., & Brooks, J. (1974). Self, other and fear: Infants' reactions to people. In M. Lewis & L. A. Rosenblum (Eds.), *Origins of behavior*. Vol. 2, *The origins of fear*.

Loftus, G., & Loftus, E. (1984). *Mind at play: The psychology of video games.* New York: Basic Books.

Lowry, P. (1993). Designing for the child's privacy in the preschool environment. *Proceedings of the Environmental Design Research Association, 24*, 180–187.

Lyons, E. (1983). Demographic correlates of landscape preference. *Environment and Behavior, 15*, 487–511.

Malone, T. W. (1981). Toward a theory of intrinsically motivating instruction. *Cognitive Science, 5*, 330–370.

Mandinach, E. B., & Corno, L. (1985). Cognitive engagement variations among students of different ability level and sex in a computer problem solving game. *Sex Roles, 13*, 241–251.

Marks, I. M. (1987). *Fears, phobias and rituals: Panic, anxiety and their disorders.* New York: Oxford University Press.

Matthews, M. H. (1992). *Making sense of place: Children's understanding of large-scale environments.* Hertfordshire, UK: Simon & Schuster International.

McNally, R. J. (1987). Preparedness and phobias: A review. *Psychological Bulletin, 101*, 283–303.

Miller, R. A. (1984). Children's use of the outdoor environment. Master's thesis, Department of Landscape Architecture, University of Washington.

Moore, G. T. (1986). Effects of the spatial definition of behavior settings on children's behavior: A quasi-experimental field study. *Journal of Environmental Psychology, 6*(3), 205–231.

Moore, R. C. (1989). Plants as play props. *Children's Environments Quarterly, 6*, 3–6.

Moore, R. C., & Young, D. (1978). Childhood outdoors: Toward a social ecology of the landscape. In I. Altman & J. F. Wohlwill (Eds.), *Children and the environment.* New York: Plenum Press.

Moran, C., & Andrews, G. (1985). The familiar occurrence of agoraphobia. *British Journal of Psychiatry, 146*, 262–267.

Morgan, G. A., & Ricciuti, H. N. (1967). Infants' responses to strangers during the first year. In B. M. Foss (Ed.), *Determinants of infant behavior.* London: Methuen.

Muris, P., Merckelbach, H., & Collaris, R. (1997). Common childhood fears and their origins. *Behavioral Research and Therapy, 35*, 929–937.

Murray, J. P., & Kippax, S. (1978). Children's social behavior in three towns with differing television experience. *Journal of Communication, 28*, 19–29.

Neel, J. V. (1970). Lessons from a "primitive" people. *Science, 170*(3960), 815–822.

Newson, E. (1996). Video violence and protection of children. In T. Gill (Ed.), *Electronic: How children are responding to the information revolution.* London: National Children's Bureau.

Öhman, A. (1986). Face the beast and fear the face: Animal and social fears as prototypes for evolutionary analyses of emotion. *Psychophysiology, 21*, 123–145.

Öhman, A., & Soares, J. J. F. (1993). On the automaticity of phobic fear: Conditioned responses to masked phobic stimuli. *Journal of Abnormal Psychology, 102*, 121–132.

Öhman, A., & Soares, J. J. F. (1994). Unconscious anxiety: Phobic responses to masked stimuli. *Journal of Abnormal Psychology, 103*, 231–240.

Okagaki, L., & Frensch, P. A. (1994). Effects of video game playing on measures of spatial performance: Gender effects in late adolescence. *Journal of Applied Development Psychology, 15*, 33–58.

Ollendick, T. H., Matson, J. L., & Heisel, W. J. (1985). Fears in children and adolescents: Normative data. *Behavior Research and Therapy, 23*, 465–467.

Olds, A. R. (1987). Designing settings for infants and toddlers. In C. S. Weinstein & T. G. Davis (Eds.), *Spaces for children: The built environment, and child development*. New York: Plenum Press.

Orians, G. H. (1980). Habitat selection: General theory and applications to human behavior. In J. S. Lockard (Ed.), *The evolution of human social behavior.* Chicago: Elsevier.

Orians, G. H., & Heerwagen, J. H. (1992). Evolved responses to landscapes. In J. Barkow, L. Cosmides & J. Tooby (Eds.), *The adapted mind: Evolutionary psychology and the generation of culture*. Oxford: Oxford University Press.

Pagel, M. D., & Harvey, P. H. (1989). Taxonomic difference in the scaling of brain weight on body weight among mammals. *Science, 244*, 1589–1593.

Pinker, S. (1997). *How the mind works*. New York: Norton.

Poulton, R., Davies, S., Menzies, R. G., Langley, J. D., & Silva, P. A. (1998). Evidence for a non-associative model of the acquisition of a fear of heights. *Behavior Research and Therapy, 36*, 537–544.

Russell, P. A. (1979). Fear-evoking stimuli. In W. Sluckin (Ed.), *Fears in animals and man*. New York: Van Nostrand Reinhold.

Scarr, S., & Salapetek, P. (1970). Patterns of fear development during infancy. *Merrill-Palmer Quarterly, 16*, 53–90.

Schaffer, H. R. (1974). Cognitive components of the infant's response to strangeness. In M. Lewis & R. Rosenblum (Eds.), *Origins of fear* (pp. 10–24). New York: Wiley.

Scherer, M. W., & Kakamura, C. Y. (1968). A fear survey for children: A factor analytic comparison with manifest anxiety. *Behavior Research and Therapy, 6*, 173–182.

Sheldon, L. (1998). The middle years: Children and television—cool or just plain boring? In S. Howard (Ed.), *Wired up: Young people and the electronic media*. London: UCL Press.

Skarin, K. (1977). Cognitive and contextual determinants of stranger fear in six- and eleven-month-old infants. *Child Development, 48*, 537–544.

Smith, P. K. (1979). The ontogeny of fear in children. In W. Sluckin (Ed.), *Fear in animals and man.* New York: Van Nostrand Reinhold.

Smith, P. K., & Soloboda, J. (1986). Individual consistency in infant-stranger encounters. *British Journal of Developmental Psychology, 4*, 83–91.

Solomon, R., & DeCarie, T. G. (1976). Fear of strangers: A developmental milestone or an overstudied phenomenon? *Canadian Journal of Behavioral Science, 8*, 351–362.

Stutz, E. (1996). Is electronic entertainment hindering children's play and social development. In T. Gill (Ed.), *Electronic children: How children are responding to the information revolution.* London: National Children's Bureau.

Ulrich, R. (1993). Biophilia and biophobia. In S. R. Kellert & E. O. Wilson (Eds.), *The biophilia hypothesis*. Washington, DC: Island Press.

Uzgiris, I. C. (1967). Ordinality in the development of schemas for relating to objects. In J. Helmuth (Ed.), *Exceptional infant-normal infant* (pp. 317–334). New York: Bruner/Mazel.

White, B. L., Kaban, B., Shapiro, B., & Attanucci, J. (1977). Competence and experience. In I. C. Uzgiris & F. Weizmann (Eds.), *The structure of experience* (pp. 15–152). New York: Plenum Press.

White, R., & Heerwagen, J. (1998). Nature and mental health. In A. Lundberg (Ed.), *The environment and mental health*. Mahwah, NJ: Erlbaum.

Williams, T. M., & Handford, A. G. (1986). Television and other leisure activities. In T. M. Williams (Ed.), *The impact of television: A natural experiment in three communities* (pp. 143–214). Orlando, FL: Academic Press.

Wilson, E. O. (1994). *Naturalist.* Washington, DC: Island Press.

Zegans, S., & Zegans, L. S. (1972). Fear of strangers in children and the orienting reaction. *Behavioral Science, 17*, 407–419.

3

The Development of Folkbiology: A Cognitive Science Perspective on Children's Understanding of the Biological World

John D. Coley, Gregg E. A. Solomon, and Patrick Shafto

Human beings are intellectually adventurous. We divide the world into kinds of things, such as living and nonliving, but we are also driven to go beyond categorization. We seek to understand and to explain. The understandings and explanations that we construct to make sense of the world can be thought of as *folk theories*. Folk theories are informal, often intuitive ways of explaining the *what* and the *why* of the world. Folk theories play a central organizing role in determining how children (or adults, for that matter) understand new facts (Gopnik & Wellman, 1994; Wellman & Gelman, 1998). We relate new information to old explanations; we understand it in terms of the theories or mental frameworks we already possess. For example, a child whose theory of the universe has the earth at its center and who believes in an absolute up and down (as opposed to a relative one) would likely explain night and day differently than we would and would probably make very different predictions about what would happen to a rock dropped into a hole dug through the earth.

One folk theory that has received attention in developmental psychology, anthropology, and the philosophy of science is *folkbiology*. Folkbiology refers to the cognitive processes by which people understand, classify, reason about, and explain the world of plants and animals. Our survival as a species has depended in large part on acquiring an intimate knowledge of plants and animals. At its core, research in folkbiology asks to what extent this dependence has shaped our basic conceptual apparatus. Do concepts like *plant*, *ferret*, and *rainbow trout* differ from concepts like *furniture*, *car*, and *ballpeen hammer*? Do we generalize about unfamiliar birds in the same way we generalize about unfamiliar lawnmowers? Do we explain catching a cold in the same way we explain catching a baseball or catching our sleeve on a sharp edge? More generally, how is thinking about the living world like or unlike thinking about other domains of experience?

Folkbiology is not taught in schools; indeed, it often clashes with what we are formally taught in biology class. It does not involve the sanctioned means of gathering evidence and testing hypotheses that mark formal science. But that is not to say that folkbiology is a simple collection of facts and beliefs. It is more than that. It is a theory, and as with any theory, folkbiology is defined in terms of the phenomena it explains and the range of entities and causal processes it entails. Folkbiology provides predictions and supports explanations about such phenomena as growth, digestion, death, and illness—the phenomena of living things. If we eat too much, we gain weight. If deprived of air, we die. These are predictions we make on the basis of folkbiological understanding.

Recent research in the development of folkbiology has revolved around two related but separable issues. The first issue is to what extent and at what point in development folkbiological thought is truly distinct from thought in other domains. Do we have specialized conceptual tools for thinking about nature, or do we reason about plants and animals much as we reason about other aspects of experience? If we do possess these specialized tools, at what point in development do they emerge? To what extent, for example, are there early arising, perhaps even innately derived, dispositions to reason about particular biological entities and phenomena in particular ways? At what age ought we deem children's biological understanding sufficiently coherent to constitute a folk theory?

The second issue is the nature of developmental change in folkbiology. Does the organization of folkbiological knowledge undergo *quantitative* change as knowledge is acquired and initial principles are elaborated, or is it more a matter of *qualitative* change as one construal of the biological world is overthrown for another? Is the understanding of nature seen in the young child simply a less elaborate version of that seen in the adult, or do children and adults possess deeply different worldviews with respect to folkbiology, necessitating radical conceptual change at some point in development?

Thus, the study of folkbiology is central to the question of how children understand nature. But it also raises a number of other vital questions about the nature of thinking in general. In this chapter, we review empirical evidence on the development of children's folkbiological thought, using these two issues to guide our review. We then offer our own synthesis of a plausible model of development and highlight what questions remain outstanding in this domain of inquiry. Even the partial answers we come up with have important implications for science education as well as for public health and environmental awareness initiatives. They also tell us something about who we understand ourselves to be.

The Nature of Early Folkbiology

In this section we review evidence to date on what children know about living things, with an eye toward addressing the two questions raised above about the uniqueness of folkbiological thought and the nature of developmental change therein. We focus on areas central to any folk theory of biology: What kinds of things are alive, what is the place of humans in the natural world, what sorts of biological causality do children understand, and what are the unique properties of biological categories?

Animism

A crucial component of any biological understanding is the ability to differentiate living from nonliving things. By definition, biology is the idea that living things are different in important ways from nonliving things. One noted aspect of early reasoning about biology is the phenomenon of childhood animism, wherein young children report (and allegedly believe) that inanimate objects are alive, a trend that contrasts strikingly with adult notions and has led researchers to posit large, qualitative differences between the biological understandings of children and adults (e.g., Piaget, 1929; Carey, 1985). Piaget (1929) asked children which of a range of entities was alive. For example, "Is the sun alive? Is a dog alive? Is a flower alive?" He found that young children did not restrict their judgments of what is alive to the ontological category *living thing* but also extended them to such inanimate objects as cars, clouds, and even statues. These findings were replicated more systematically by Laurendeau and Pinard (1962) and suggest that the basis for children's decisions about what is alive and therefore a legitimately biological object is very different from that used by adults.

However, more recent evidence suggests that early studies overestimated animistic reasoning. For example, Richards and Siegler (1984) systematically asked children ages four through 11 whether various objects (people, animals, plants, vehicles, other inanimate objects) that were described as either being still, being moved, or (where plausible) moving themselves were alive. Of interest was whether over the entire set of questions children's responses corresponded to systematic rules. Results showed that children rarely attributed life to vehicles and objects and never did so systematically. Most younger children systematically attributed life to people and animals, and by around age eight most children had added plants. Thus the largest developmental shift was not learning that inanimates are not alive but learning that plants are alive (see also Carey, 1985; Dolgin & Behrend, 1984; Richards, 1989).

Evidence also suggests cultural and experiential differences in patterns of life judgments. Hatano, Siegler, Richards, Inagaki, Stavy, and Wax (1993) present data showing that Japanese children may be more liberal than U.S. children in granting life status to objects such as mountains and that Israeli children are more conservative, often denying even that plants are alive. These findings are tied to specific beliefs in each culture. Recent evidence further suggests that urban, rural, and Native American children in the United States may differ in willingness to attribute life to plants. Ross, Medin, Coley & Atran (under review) find that all three of these groups are at ceiling, attributing life to an assortment of animals, ranging from bears to worms. However, Native American children are most likely to affirm that plants are alive. Rural children are less likely but show a developmental increase in the attribution of life to plants, and urban children tended neither to attribute life to plants nor to show such development.

Although they may not explicitly identify plants as living things, preschoolers may nevertheless understand important commonalities between plants and animals. For example, four-year-olds reliably report that plants and animals, but not human-made artifacts, can spontaneously heal or regrow injured parts (Backscheider, Shatz & Gelman, 1993). Four-year-olds also show a clear understanding of seeds and plant growth and of the underlying similarities between growth in plants and in animals (Hickling & Gelman, 1995). Inagaki and Hatano (1996) present evidence that five-year-olds projected biological properties such as growing and needing water to both plants and animals and coherently explained biological processes (taking in nutrients, growth, death) for plants by drawing on analogous properties for animals. Thus, the knowledge that plants are like animals and unlike nonliving things in important ways may be implicit in young children's reasoning.

In sum, contrary to classic Piagetian research, more recent and systematic investigations reveal little evidence of childhood animism. Most preschoolers systematically report that animals (whether mammals, fish, or bugs) are alive and that inanimate objects (whether bicycles or pencils) are not. Beliefs about the status of plants are less clear-cut. When asked directly, children are less consistent in reporting that plants are alive. However, children report that plants, like animals, grow, need water and air, and die, thereby acknowledging important biological commonalities between plants and animals. Finally, beliefs about the status of plants seem susceptible to cultural influences.

The finding that preschoolers do not consider plants to be alive suggests a qualitative difference between children and adults with respect to a core biological concept. Children might take behavior as their metric of living things and humans as the prototypical behaving (hence living) thing. However, when questioned more closely, preschoolers affirm that plants participate in the same basic biological processes as animals. Like adults, they appear to have carved out the domain of living things—including plants—as being united on the basis of important biological concepts. This early understanding may be frail and may not immediately be linked with the term *alive*, but it appears more compatible with a model espousing quantitative developmental change. Ideas of what is alive seem not to undergo radical revision with development.

Anthropocentrism

Carey (1985, 1995) has argued that children's early understanding of plants and animals is anthropocentric. In other words, children's understanding of other living things is largely in reference to, or by analogy to, human beings (see also Inagaki & Hatano, 1991). Moreover, if young children see humans as the prototypical living thing, they should reason about other animals and plants on the basis of similarity to humans rather than on principles of biological necessity.

Several lines of evidence fit this characterization. Carey asked children which of a set of biological properties (such as breathes, eats, has bones, has babies) they believed could be attributed to a series of entities, ranging from humans to animals to plants to inanimate objects. For example, she asked, "Do dogs have baby dogs?" Results suggest that children attributed properties on the basis of similarity to humans (but see Coley, 1995). Analogous patterns were observed when children were taught a new fact about a given biological kind (for example, a dog "has an omentum") and asked whether other kinds (a bird, a fish, a plant) share that property. Carey (1985) reports a pattern of results consistent with the view that four- and six-year-old children's conceptions of the natural world are indeed anthropocentric. First, overall projections from humans were stronger than projections from other living things. Second, specific asymmetries in projection emerged, such that (for example) inferences from human to dog were stronger than from dog to human. Finally, children's reasoning followed striking violations of similarity, such that (for example) inferences from human to bug were stronger than from bee to bug. These patterns suggest that human is a privileged inferential base for the children Carey studied.

Carey (1985) interpreted her results from these and other tasks as indicating that young children's earliest reasoning about biological phenomena is not biological per se. She argues that these reasoning patterns reflect the operation of general—rather than specifically biological—reasoning mechanisms. Biological properties are attributed on the basis of comparison to a central exemplar—humans—rather than on the basis of membership in the class of living things, independent of similarity to humans. Preschoolers, according to Carey, lack knowledge of basic biological causal mechanisms and so have not yet constructed a naive theory of biology that can provide explanations and support predictions of phenomena. Rather, their understanding of living things centers around behavior and human beings as the prototypical behaving being. This pattern of reasoning has been interpreted as demonstrating that young children possess an understanding of biological phenomenona incommensurate with that of adults and that pervasive conceptual change is necessary for children to acquire the adult model in which humans are seen as one animal among many.

However, it is important to examine the generality of this anthropocentric pattern of reasoning on at least two grounds. First, rather than being diagnostic of deep conceptual commitments, this anthropocentric folkbiology may reflect a lack of knowledge about the biological world. Carey's subject population, in Cambridge, Massachusetts, may be relative folkbiological novices. Some evidence suggests that children who are more familiar with certain living kinds prefer to use knowledge of those kinds in reasoning. Inagaki (1990) showed that children who raised goldfish reasoned about a novel aquatic animal (a frog) by analogy to the goldfish, not to humans. So perhaps Carey's population (and that studied by most developmental researchers) did not have sufficient knowledge of nonhuman living kinds to use them as an inferential base. Increased knowledge might provide more salient biological exemplars that could in turn mitigate anthropocentrism.

Second, an anthropocentric folkbiology may reflect cultural assumptions about relations between humans and nature. Again, in the population studied by Carey, the differences between humans and nonhumans are very sharply drawn. Direct interaction with and dependence on nature is relatively rare. In urban, industrialized Western societies, humans are seen as existing apart from nature. In a culture where humans are perceived as an integral part of nature, people might be less likely to make anthropocentric construals.

In an ongoing comparative study of members of the Menominee Indian tribe of Wisconsin, Coley and his colleagues (Ross et al., under review) are currently addressing some of these questions. The Menominee are interesting for a number of reasons. First, according to the traditional Native American view, humans are an integral part of the natural world (Bierhorst, 1994; Suzuki & Knudtson, 1992). This contrasts sharply with the predominant Western view. Second, traditional folkbiological knowledge is especially salient to the Menominee. Unlike many woodland tribes, the Menominee reservation occupies (a small fraction of) their traditional range; thus, traditional knowledge of local plant and animal species is still very relevant today. Moreover, the Menominee run a successful logging operation that employs traditional ecological knowledge to guide forest management. Finally, Menominee children differ from a typical urban or suburban sample in terms of both having a cultural tradition of viewing humans as an integral part of the natural world and having a great deal of experience with plants and animals. Children spend time fishing and hunting and in general have a high degree of contact with plants and animals. And indeed, contrary to results with middle-class urban children, Menominee children ages six and above show no evidence of anthropocentric folkbiological reasoning (Coley, Medin & James, 1999). A property projection task like that used by Carey revealed no evidence that *human* functions as a privileged inductive base, little evidence of asymmetries in projections, and no evidence for violations of similarity. Rather, Menominee children's projections were based largely on similarity among living things and to some extent on causal and ecological relations (see also Ross et al., under review). It appears that early folkbiology is neither universally nor inevitably anthropocentric.

In sum, most urban children have relatively little interactive experience with a range of living things and little cultural support to see humans as one living thing among many. For them, humans may initially provide a salient exemplar for biological reasoning. For populations with different levels of experience or cultural beliefs, this anthropocentric perspective appears to be strongly mitigated. Nor does attribution of life to animals seem based on comparison to a human exemplar. We currently lack information on the degree to which urban adults' folkbiology is anthropocentric and so cannot assess the degree to which children's views are discontinuous. But it would appear that anthropocentric folkbiology is not an inevitable step in development and that concentration on a single population (urban or suburban children from industrialized societies) may overemphasize discontinuity in development. More generally, these findings suggest that the very model of development within the domain of folkbiology may vary for different populations.

Biological Causality

As stated above, a central component of having a folk theory is having an understanding of the kinds of causes that unify facts falling under the scope of that theory. Having a biological theory entails having an understanding of uniquely biological causal mechanisms. Do children have an understanding of the causal principles that govern biological change? To answer this question, researchers have examined preschoolers' understandings of specific phenomena such as inheritance, illness, and growth, with an eye toward how they organize facts to provide uniquely biological explanations and prediction.

Inheritance Biological inheritance has emerged as a central phenomenon in the debate over the development of domain-specific explanatory theories, for our adult understanding of it requires the interrelation of a system of concepts by means of core causal principles. At a commonsense level, an understanding of biological inheritance entails understanding and causally interrelating at least three concepts—that offspring tend to resemble their parents, that this resemblance pertains principally to intrinsic biological rather than acquired psychological traits, and that such resemblance is fixed by mechanisms eventually culminating in birth. Thus, if a daughter resembles her blonde mother because she has bleached her hair, we would not say that such a resemblance is an example of biological inheritance, nor would we say that a child's knowing, like his mother, where the cookie jar is hidden is an example of biological inheritance.

Researchers have shown that preschoolers know many, if not all, of the separate facts entailed in such an understanding. Recent studies have convincingly demonstrated that preschoolers understand that bodily traits and mental traits are different sorts of things and therefore that bodily traits lie outside of the explanatory purview of a naive theory of psychology (e.g., Inagaki & Hatano, 1993; Kalish, 1997). Preschoolers know, for example, that desiring and learning cannot directly alter bodily features as they can alter mental features; you cannot grow a third eye simply by thinking about it. Second, there is evidence that preschoolers understand that offspring will tend to resemble their parents: dogs tend to have puppies and not kittens; people with dark skin tend to have children with dark skin and not light skin (Gelman & Wellman, 1991; Hirschfeld, 1995; Johnson & Solomon, 1997; Springer & Keil, 1989). And third, many, if not most, preschoolers know that babies come from their mothers' bellies (Bernstein & Cowan, 1975; Springer, 1995).

Despite an impressive knowledge of these facts about inheritance, recent work suggests that many preschoolers make judgments that are not consistent with the commonsense understanding of inheritance outlined above. The question at hand is whether children conceptually interrelate these facts in terms of underlying biological causal mechanisms. In a series of recent studies (Hirschfeld, 1995; Johnson & Solomon, 1997; Solomon, 1996; Solomon, Johnson, Zaitchik & Carey, 1996; Springer, 1996; Weissman & Kalish, 1998), preschoolers were asked to judge whether adopted children would be more likely to resemble their birth parents or their adoptive parents on a range of traits. Most preschoolers did not show the adultlike pattern of judgment that children should resemble their birth parents on inborn physical traits and their adopted parents on acquired beliefs. Thus, although they know the facts of birth and distinguish mental from physical properties, most preschoolers appear to have difficulty coordinating this knowledge in a coherent fashion. Note, however, that when the task is simplified, a significant minority of preschoolers do perform as adults do. Moreover, many of those children who do not show an adultlike performance on the task still appear to think that the birth parent has special status in regard to parent-child resemblance, though they appear not to have completely worked out the implications of that relationship.

Thus, on the one hand, preschoolers' reasoning about the phenomenon of biological inheritance would appear to undermine the broad claim that they have an understanding of folkbiology consistent with that of adults and so has been claimed as support for a qualitative change model of the acquisition of folkbiology (Carey, 1995; Solomon et al. 1996). But on the other hand, some researchers argue that the simple undifferentiated bias of many preschoolers to regard birth parentage as special is itself an indication that children have the rudiments of a causal understanding of inheritance. They argue that such an association indicates that preschoolers understand there to be some implicit biological mechanism fixing parent-child resemblance, though their understanding of the phenomenon must undergo further refinement (e.g., Gelman & Hirschfeld, 1999; though see Solomon, 1996). These researchers take children's understanding of the phenomenon as support for the claim that folkbiology is a domain that undergoes quantitative change. Suffice it to say that an understanding of biological inheritance consistent with that of Western biology is, at least to some extent, an acquired understanding and, for preschoolers, a fairly fragile one at that. Of course, even if it were shown that preschoolers do not reason about inheritance in terms of uniquely biological processes or that this birth-parent bias is not universal, this would not preclude preschoolers from reasoning in terms of biological causes about other phenomena. Indeed, because of the sophisticated interrelation of concepts that is required to understand it, biological inheritance may simply be an unlikely candidate for inclusion in preschoolers' early folkbiology. Preschoolers' folkbiology may not yet include all of the phenomena that adult folkbiology does.

Illness For adults, illness—like inheritance—is a biological process. Our adult folk theory of germ-based illness can be said to comprise three core concepts. Children must conceptualize the facts regarding contamination, contagion, and symptoms and interrelate them as adults do for a functioning theory of illness. For example, adults would attest that a person cannot catch a cold from another person by talking to him or her over the phone but could catch a cold by using a phone directly after it was used by a person with a cold.

There is evidence that preschoolers know many of the facts that we adults know to be relevant to a germ theory of illness. Certainly, preschoolers know that some things are bad for you. And they also know that there is such a thing as contamination and that contaminants may be invisible. For example, preschoolers will consider a beverage to be undesirable if it has had a cockroach or feces placed in it, even after the contaminant has been removed. And they will predict that drinking such contaminated beverages can make you ill (Rozin, Fallon & Augustini-Ziskind, 1985). Preschoolers also know the facts of contagion; they know that certain symptoms are contagious whereas others are not. They know, for example, that you can catch a cold but not a scraped knee from another person. And, as Kalish (1996a, 1996b) has demonstrated, preschoolers understand that contact with germs can make you ill.

Preschoolers may know most of the facts relevant to a germ theory of illness. They hear such commands as "Watch out, I don't want you to catch my cough" and so learn that coughs are the sorts of things that can be caught; and they hear "Don't eat that, it has germs on it, and germs will make you sick." The question is whether children recruit these facts as part of an interrelated system of causal explanation. As adults, we understand that a cold is contagious but a scraped knee is not because of what we infer about how the symptom was acquired and what therefore underlies the symptom. We link our understandings that germs can cause colds to our understandings that colds can be contagious. But do preschoolers? It could simply be that preschoolers know that certain symptoms are contagious but never understand the relevance of the acquisition of the symptom by germs.

A recent study highlights this distinction. Solomon and Cassimatis (1999) designed a series of tasks to determine if children could link their knowledge of the acquisition of a symptom to its subsequent spread. Characters were described as having particular symptoms, such as runny noses or belly aches. In some conditions, the symptom was described as having been caused through contact with germs. For example, "A girl named Sandy breathed in some germs, and pretty soon she got a runny nose and had to stay home from school." In other versions, the cause of the symptom was attributed to a particular event involving an irritant. For example, "A girl named Sandy breathed in some pepper, and pretty soon she got a runny nose and had to stay home from school." The children were then asked whether they thought that the symptom could be caught by a friend who played with the ill child.

Adults, as is consistent with their having a germ theory of illness, judged the symptoms to be contagious if they were originally caused by germs but not contagious if they were caused by the irritant events. By contrast, not even half of the children under the age of 10 years made this distinction. Preschoolers, for example, judged the symptoms caused by germs to be contagious 84 percent of the time but also judged symptoms caused by events such as smoking to be contagious 72 percent of the time (the difference was not statistically significant). Preschoolers appear not to differentiate germs as contagious biological disease agents from nonbiological and therefore noncontagious symptom-causing agents such as poisons. In short, the preschoolers did not provide evidence that they understand germs to be a core explanatory concept in their understanding of illness, one that brings their separate understandings of contamination, contagion, and symptoms into contact with one another.

Preschool children appear not to have an adultlike understanding of illness and contagion. Children do not interrelate contamination and contagion with differences between germs and poisons, nor do they identify germs as living things. How do we characterize this movement to an adultlike understanding in the context of a folkbiological theory? If differentiation of biological from nonbiological causal agents is considered central, then the movement to an adultlike germ theory might be considered qualitative change as children focus on what is properly deemed biological. However, these lines of evidence are not at odds with a quantitative model of folkbiological development at large. Perhaps children have a foundational understanding of contagion as a biological process in place early, and facts about germs are simply learned later in development and tacked onto this preexisting folkbiological theory of contagion. In other words, germs may be a fringe case of folkbiology, a detail rather than a central fact (see also Keil, Levin, Richman & Gutheil, 1999).

Growth and Natural Change Another central component of folkbiology is the idea that living things spontaneously grow and change, whereas other kinds of things do not. What do children know about growth and natural change? And more important, is there evidence that this knowledge is rendered coherent by a framework of causal principles?

Inagaki and Hatano (1996) present evidence that four- and five-year-olds believe that animals and plants, but not artifacts, spontaneously change over time. Rosengren, Gelman, Kalish, and McCormick (1991) showed that three- and four-year-olds understand that animals grow over time. This belief in the power of growth over time even led preschoolers to make nonnormative predictions; when presented with a picture of a small caterpillar and asked whether it would grow up to be a large caterpillar or a butterfly, children chose the former. They were not just reporting what they had observed but using beliefs about growth to make (in this case, incorrect) inferences.

While the extrapolation of size increases over time to nonnormative cases is a potent example of three-year-olds' understanding of biological growth, to truly possess an adultlike understanding of growth requires both the prediction that animals grow as a function of time and the prediction that artifacts do not grow as a function of time. Rosengren and his colleagues (1991) investigated this matter with interesting results. When three-year-olds were presented with an artifact and asked whether it would be the same size or larger after a period of time, they responded at chance levels. This would suggest that these children were unable to restrict their generalizations of growth to just animals. However, closer inspection of response patterns revealed order effects. When children were presented with animals before artifacts, they were more likely to apply the growth model to the artifacts. When artifacts were presented first, there was no evidence of carryover of a nongrowth model to animals. These results suggest that biological growth is an established model at three years of age but that its restriction to the biological domain is fragile.

Studies of children's understanding of growth and natural change favor a quantitative model of conceptual change. In these studies, preschoolers—like adults—report that animals and plants grow over time. Indeed, they overgeneralize this prediction to apply to species that undergo metamorphosis, a hallmark of theory-based reasoning. This understanding of growth seems fragilely biological; in some cases these predictions are restricted to living things and in some cases they are not.

Structure of Plant and Animal Categories

A major task facing the young child is to divide the world into discrete classes of things. By classifying objects, we can better understand them and make predictions about individuals that we have never seen before (Medin & Coley, 1998). There is evidence that children's categories of living things may have unique structural properties when compared to categories in other domains, such as inanimate objects (e.g., Gelman & Coley, 1991). Here we briefly touch on two ways in which children's living kinds may be structured differently from other concepts: children appear to assume that members of a living kind share an essence and that category membership is a reliable guide for inferences about shared underlying properties.

Essentialism Psychological essentialism (Medin & Ortony, 1989) asserts that people act as if there is an underlying property that caused the observable characteristics of any item. It is this property (real or imagined), not observable characteristics, that ultimately determines category membership. For example, by age 10, children believe that a raccoon painted black with a white stripe and with a pouch of "smelly stuff" remains a raccoon despite its outward similarity to a typical skunk (Keil, 1989; see also Gelman, 2000, for a review of this evidence). This suggests that for 10-year-olds, something other than outward appearance makes an object a skunk. Two questions are of interest to the present discussion: at what age do children begin to show such essentialist reasoning tendencies, and are these reasoning patterns specific to folk-biological kinds, or do they reflect domain-general reasoning?

One consequence of essentialist beliefs is that superficial transformations should not change the identity of a living thing. Keil (1989) told kindergartners, second-graders, and fourth-graders stories about how animals or objects of one kind were altered to resemble animals or objects of another kind (for example, a raccoon was painted to resemble a skunk, or a coffeepot was made into a birdfeeder). When queried about artifacts, Keil found children of all ages believed that transformations changed identity. For natural kinds, there was a developmental progression: older children denied that transformations changed the identity of the object; a raccoon painted to look like a skunk was still a raccoon. Younger children rejected transformations that crossed ontological boundaries, judging that a porcupine, for example, cannot be turned into a cactus. In contrast, young children relied more on surface appearances to make decisions about category membership after other transformations, stating, for instance, that the painted raccoon was indeed a skunk. This suggests that although they may have some rudiments of an essentialist assumption in place, younger children would appear not to have the same biological understanding of the immutability and origins of species identity as do adults (see also Johnson & Solomon, 1997; Keil, 1994). In addition to having portions of the essentialist bias in place with respect to plants and animals, by second grade children clearly differentiate between the results of transformations on kind membership for biological entities and artifact kinds. The children say that the same kind of superficial transformations that do not change category membership for plants and animals do change category membership for artifacts. This suggests that folkbiological kinds are subject to different causal laws than artifact kinds.

Another consequence of essentialism is the belief that a member of a kind has the innate potential to become like other members of the kind. Gelman and Wellman (1991), for example, have shown that preschoolers as young as four-year-olds believe that a baby animal would come to acquire the features characteristic of animals of its type whether or not it was raised with others of the same type, simply by virtue of being of its kind. Thus, a baby cow, raised by pigs, would come to moo and not oink because its essence is that of a cow, not of a pig. Even if planted with flowers, a seed from an apple would nevertheless grow into an apple tree. This is particularly remarkable because in the case of the seed there is no clear perceptual link between the seed and the adult plant, yet children believe that the origin of the seed, an apple, will determine what kind of plant it will eventually grow into. Even three-year-olds showed evidence that they discerned between the importance of insides and outsides in kind membership, for both animals and complex artifacts. Taken together, these results suggest that children have essentialist reasoning biases from early in development and that over time these biases become more pronounced for living kinds than for artifact categories.

Inductive Potential Related to the idea that members of the same category share deep underlying properties is the notion that categories guide inductive inferences. For instance, if asked to infer anything about *Bellin*, you would probably be at a loss. However, were I to mention that Bellin is a *cat*, you could then infer a great deal about him: he eats meat, meows, likes to be scratched between his ears, brings home dead mice, and so on. Thus, knowing the category membership of an object licenses inductive inferences about that object. One question that has arisen within folkbiology is whether living kind categories are particularly privileged with respect to inductive potential.

Experimental evidence clearly shows that children as young as two-year-olds assume that members of named kinds share underlying properties despite superficial dissimilarities (e.g., Gelman & Coley, 1990; Gelman & Markman, 1986, 1987). For example, Gelman and Markman (1986) showed young children pictures of two animals (such as a brontosaurus and a rhinoceros), gave category names to the animals (dinosaur and rhinoceros), and told them that each had a particular nonobvious property (cold-blooded or hot-blooded). The children were then shown a third animal (a triceratops) more superficially similar to the rhino but given the same category name as the brontosaurus. When asked which property it would be more likely to have, children made their inference on the basis of kind membership rather than superficial similarity. Despite looking like the rhino, the triceratops would share underlying properties with the other dinosaur. Numerous control studies show that inferences are based on shared category membership and not merely shared labels and that category membership guides inferences about intrinsic but not superficial properties.

These results show that very young children already assume that members of a named kind are likely to share novel features. Is this a general assumption about all categories, or is it limited to living kinds? Evidence to date supports the latter; there is some evidence that preschoolers are more willing to generalize within natural kind than artifact categories (Gelman & O'Reilly, 1988) and that by age eight children clearly and consistently show stronger inductions within natural kinds than artifact categories (Gelman, 1988; Gelman & O'Reilly, 1988). Note that this research has not explicitly examined generalizations within living kinds versus nonliving natural kinds (such as water and salt). Nevertheless, results suggest that an emerging component of folkbiological knowledge is the assumption that folkbiological categories are stronger than artifact categories as guides to inductive inference. In other words, by age two and a half categories reliably guide children's inductive inferences, and by age eight children, like adults, are more likely to bet that living kinds are alike with respect to novel underlying features than are other kinds of objects in the world.

However, children's category-based inductions do differ from those of adults. The similarity-coverage model of Osherson, Smith, Wilkie, López, and Shafir (1990) predicts that an argument whose premises are more diverse will be judged stronger than an argument whose premises are similar. For example, one should be more willing to generalize to all birds from sparrows and flamingos than from sparrows and robins. Osherson et al. attributed this finding to *coverage*. Reasoners compare the taxonomic similarity of the premise categories (such as sparrows and flamingos) to sampled members of the more general conclusion category (kinds of birds). The premise set with better coverage—that is, higher taxonomic similarity to sampled instances—makes for the stronger argument. For the most part, undergraduate research subjects reason in accordance with this diversity principle, based on taxonomic similarity. However, developmental studies reveal that children through age 10 have difficulty grasping this phenomenon when reasoning about living things (Lopez, Gutheil, Gelman & Smith, 1992; Gutheil & Gelman, 1997; but see Heit & Hahn, 1999, for evidence that by age five children can successfully reason according to diversity when reasoning involves familiar properties).

Research in this area suggests similarities and differences between category-based induction for adults and children. Both show clear appreciation for the inductive potential of categories, but certain inductive phenomena seen among some adult populations (such as diversity-based reasoning) are hard to find among children. What is most at question is whether these characteristic induction patterns are particular to biological reasoning—perhaps tied to the discovery of higher-order biological categories (such as *mammal* or *living thing*)—or whether they represent a more general cognitive mechanism. Certainly nonliving kinds license inferences, so to some extent inductive potential must be a general cognitive mechanism. Is inductive potential especially strong in living kind categories? What scant evidence there is supports the case, but more research is needed. Current evidence does not suggest strong discontinuities in induction between children and adults and does not indicate whether induction over biological kinds is special.

How to Characterize Development of Folkbiology

Researchers in this field by and large agree on what children say when reasoning about plants and animals. What remains in dispute is what the facts of early folkbiological reasoning imply about the underlying organization of knowledge and how that organization changes with development. At the beginning of the chapter we raised two related questions about the development of folkbiology. First, is the acquisition of folkbiological knowledge a relatively continuous process as knowledge is acquired and initial principles are elaborated, or is it a more discontinuous process as one construal of the biological world is overthrown for another? And second, do young children organize their knowledge of plants and animals differently from how they organize their knowledge about other kinds of objects?

We believe that evidence available to date supports a mixed model of development. Clearly, by the time they begin formal education, children use nature as a salient and distinct domain of inquiry as they try to construct an understanding of their world. Preschoolers appear to reason about the categories of plants and animals largely as adults do. Like adults, they assume that living kinds have essences and support induction. They do not differ systematically with respect to what kinds of things are considered alive. Thus, there seem to be important continuities between young children and adults in the way that folkbiological categories are structured and used in reasoning.

Does this mean that development does not occur? Absolutely not. As they grow and explore the world, children acquire vast amounts of information about plants and animals and about linking that information together in new ways. This knowledge impacts patterns of reasoning. Moreover, children appear to undergo important conceptual changes over the early school years regarding reasoning about biological causality. Evidence suggests that although children share many elements of adults' understandings of inheritance, illness, and growth, these elements are initially understood in isolation and are interrelated only in a causally coherent fashion over time. Causally coherent folkbiological principles may not be present early on but may emerge over development, suggesting discontinuities between children and adults in terms of biological explanation and therefore indicating the presence of genuine conceptual change.

The development of folkbiological thought involves both continuity and discontinuity. The way children form categories of plants and animals and the assumptions they make about the structure and functions of those categories are much like adults. However, children differ markedly from adults in the way they understand biological causality and explain biological processes. What is less clear is whether folkbiological cognition is distinct from thinking in other domains. Part of the difficulty in resolving this issue lies in our current lack of knowledge in three areas.

Specifying the Adult Endstate

An important question that has been largely implicit throughout our discussion thus far is what adult theory of folkbiology children are acquiring. To what degree to adults possess detailed knowledge of biological causal mechanisms? It is difficult to characterize the process of development without a detailed look at the adult endstate (Coley, 2000). Adult biological knowledge may be much more fragmentary and incoherent than is often assumed (Au & Romo, 1999; Keil, Levin, Richman & Gutheil, 1999). It would seem unfair to hold children to a standard seldom met by adults in their culture. Ironically, a complete understanding of the nature of developmental change of those understandings may await more detailed evidence about adult folk understanding in this domain.

Reasoning in Different Domains

Another recurrent obstacle to our attempt to assess the uniqueness of folkbiological thought is the lack of evidence comparing reasoning about living things with reasoning about other kinds of things. We are accumulating a good deal of research about how children understand and reason about living things, but without detailed studies of comparable reasoning about nonliving things it is difficult to assess whether this understanding is properly thought of as folkbiological or whether it reflects more general conceptual mechanisms. A final answer on whether and when folkbiological thought is domain specific awaits such evidence.

Folkbiological Reasoning in Cross-Cultural Perspective

A final factor that has not been adequately explored is the differences in folkbiological theories and reasoning among different cultural groups.

Coley (2000) argues that both cultural beliefs about plants and animals and amount of practical experience with plants and animals could well lead to large differences in both the endpoint of development (what do most adults in the community know?) and the path by which development proceeds (direct teaching, apprenticeship, or guided participation). At the very least, Coley, Medin, and James's (1999) results cited above suggest that cultural beliefs may affect the rate of change in children's developing theories. Bloch, Solomon, and Carey (2001) found much the same thing in a cross-cultural study of reasoning about inheritance. But as discussed above, comparative results also suggest the possibility of even greater divergences. For example, it is possible that Menominee children younger than those studied would, unlike their older siblings, look just like similar-age majority-culture children, but it is also possible that they never do. It is not clear to what extent humans are understood to be integrated into the natural world. Would the Menominee children in the Keil task judge that an animal could be transformed into a human, thereby indicating that they did not recognize the ontological boundary between humans and other animals that is so intuitive to us? By extension, would they also be more apt than we to attribute higher thought to animals? Are the strong, almost theological, reactions that the attribution of higher thought to animals draws in the West particular to a time and a place? These are empirical questions, requiring cross-cultural research involving the interrelation of complex systems of reasoning about the psychological and biological worlds.

Conclusion

We began this chapter by introducing the idea of folk theories, arguing that to understand how we organize knowledge, we must understand how that knowledge is recruited to explain different aspects of our experience. Folk theories are commonsense, informal explanations of a set of related phenomena. We then used this idea to explore how children organize their knowledge of plants and animals, known as folkbiology, focusing on the nature of developmental change and the degree to which folkbiological thought is unique. We concluded that young children look much like adults with respect to the categorization of the world into living versus nonliving things and with respect to how those categories inform inductive reasoning. In these areas, the acquisition of folkbiology seems to be a matter of quantitative change. However, children seem to undergo true conceptual change with respect to the coherent causal explanations they construct to explain biological phenomena like contagion and inheritance. It is less clear how unique folkbiological thought is, largely because the research remains to be done. More generally, by taking seriously the proposal that knowledge is organized in terms of folk theories, cognitive scientists studying the development of folkbiological thought have revealed both striking parallels in the way children and adults reason about nature and stark contrasts in the coherence of their folkbiological explanatory systems.

These recent advances in our understanding of the development of children's folkbiology have tremendous implications for science education. The implicit and explicit assumptions about the nature of cognitive development that underlie science curricula must take into account the notion that school-age children have the ability to contemplate nature with theoretical profundity. Leaving aside for now the academic debate over whether children's theories ought to be considered biological theories proper, cognitive scientists broadly agree that children bring to the classroom theories or interpretive frameworks that they use to make sense of the facts they encounter. When children's intuitive theories are consistent with the formal biology that is the target of classroom instruction, then children will be more likely to retain and integrate the key points of a class lesson. But their intuitive reasoning about biology is often not consonant with the target formal biology. For example, children's essentialist biases would appear to dispose them to understand that humans are qualitatively different kinds of beings than are other animal species. This bias to think in terms of discrete essences may well make it more difficult for children to grasp aspects of formal evolutionary theory. Of course, children's natural proclivity for theoretical reasoning also provides educators with the opportunity to engage them in explorations of the natural world at a far more profound and theoretically sophisticated level than had previously been thought possible. Just how these possibilities can best be realized is the subject of some very promising current research in the schools.

References

Atran, S. (1999). Itzaj Maya folkbiological taxonomy: Cognitive universals and cultural particulars. In D. L. Medin & S. Atran (Eds.), *Folkbiology* (pp. 119–204). Cambridge, MA: MIT Press.

Au, T. K., & Romo, L. F. (1999). Mechanical causality in children's "folkbiology." In D. L. Medin & S. Atran (Eds.), *Folkbiology* (pp. 355–402). Cambridge, MA: MIT Press.

Backscheider, A. B., Shatz, M., & Gelman, S. A. (1993). Preschoolers' ability to distinguish living kinds as a function of regrowth. *Child Development, 64*, 1242–1257.

Bernstein, A. C., & Cowan, P. A. (1975). Children's concepts of how people get babies. *Child Development, 46*(1), 77–91.

Bierhorst, J. (1994). *The way of the earth: Native America and the environment.* New York: Morrow.

Bloch, M., Solomon, G. E. A., & Carey, S. (2001). An understanding of what is passed on from parents to children: A cross-cultural investigation. *Journal of Cognition and Culture, 1*(1).

Carey, S. (1985). *Conceptual change in childhood.* Cambridge, MA: MIT Press.

Carey, S. (1995). On the origin of causal understanding. In D. Sperber, D. Premack & A. J. Premack (Eds.), *Causal cognition: A multidisciplinary debate* (pp. 268–308). New York: Oxford University Press.

Carey, S. (1999). Sources of conceptual change. In E. Scholnick, K. Nelson, S. Gelman & P. Miller (Eds.), *Conceptual development: Piaget's legacy* (pp. 293–326). Mahwah, NJ: Erlbaum.

Coley, J. D. (1995). Emerging differentiation of folkbiology and folkpsychology: Attributions of biological and psychological properties to living things. *Child Development, 66*, 1856–1874.

Coley, J. D. (2000). On the importance of comparative research: The case of folkbiology. *Child Development, 71*(1), 82–90.

Coley, J. D., Medin, D. L., & James, L. (1999). Folk biological induction among Native American children. Paper presented at the Biennial Meetings of the Society for Research in Child Development, Albuquerque, NM, April.

Diesendruck, G., & Gelman, S. A. (1999). Domain differences in absolute judgments of category membership: Evidence for an essentialist account of categorization. *Psychonomic Bulletin and Review, 6*, 338–346.

Dolgin, K. G., & Behrend, D. A. (1984). Children's knowledge about animates and inanimates. *Child Development, 55*(4), 1646–1650.

Gelman, S. A. (1988). The development of induction within natural kind and artifact categories. *Cognitive Psychology, 20*, 65–95.

Gelman, S. A. (2000). The role of essentialism in children's concepts. In H. W. Reese (Ed.), *Advances in child development and behavior* (vol. 27, pp. 55–98). San Diego: Academic Press.

Gelman, S. A., & Coley, J. D. (1990). The importance of knowing a dodo is a bird: Categories and inferences in two-year-old children. *Developmental Psychology, 26*, 796–804.

Gelman, S. A., & Coley, J. D. (1991). Language and categorization: The acquisition of natural kind terms. In S. A. Gelman & J. P. Byrnes (Eds.), *Perspectives on language and thought: Interrelations in development* (pp. 146–196). Cambridge: Cambridge University Press.

Gelman, S. A., Coley, J. D., & Gottfried, G. M. (1994). Essentialist beliefs in children: The acquisition of concepts and theories. In L. W. Hirschfeld & S. A. Gelman (Eds.), *Mapping the mind: Domain specificity in cognition and culture* (pp. 341–365). Cambridge: Cambridge University Press.

Gelman, S. A., & Hirschfeld, L. A. (1999). In D. L. Medin & S. Atran (Eds.), *Folkbiology* (pp. 403–446). Cambridge, MA: MIT Press.

Gelman, S. A., & Markman, E. M. (1986). Categories and induction in young children. *Cognition, 23*, 183–209.

Gelman, S. A., & Markman, E. M. (1987). Young children's inductions from natural kinds: The role of categories and appearances. *Child Development, 58*, 1532–1541.

Gelman, S. A., & O'Reilly, A. W. (1988). Children's inductive inferences within superordinate categories: The role of language and category structure. *Child Development, 59*, 876–887.

Gelman, S. A., & Wellman H. M. (1991). Insides and essences: Early understandings of the non-obvious. *Cognition, 38*, 213–244.

Gopnik, A., & Wellman, H. M. (1994). The theory theory. In L. A. Hirschfeld & S. A. Gelman (Eds.), *Mapping the mind: Domain specificity in cognition and culture* (pp. 257–293). Cambridge: Cambridge University Press.

Gutheil, G., & Gelman, S. A. (1997). Children's use of sample size and diversity information within basic-level categories. *Journal of Experimental Child Psychology, 64*(2), 154–179.

Hatano, G., & Inagaki, K. (1999). A developmental perspective on informal biology. In D. L. Medin & S. Atran (Eds.), *Folkbiology* (pp. 321–354). Cambridge, MA: MIT Press.

Hatano, G., Siegler, R. S., Richards, D. D., Inagaki, K., Stavy, R., & Wax, N. (1993). The development of biological knowledge: A multinational study. *Cognitive Development, 8*, 47–62.

Heit, E., & Hahn, U. (1999). Diversity-based reasoning in children age five to eight. *Proceedings of the twenty-first Annual Conference of the Cognitive Science Society* (pp. 212–217). Hillsdale, NJ: Erlbaum.

Hickling, A. K., & Gelman, S. A. (1995). How does your garden grow? Early conceptualization of seeds and their place in the plant growth cycle. *Child Development, 66*, 856–876.

Hirschfeld, L. A. (1995). Do children have a theory of race? *Cognition, 54*, 209–252.

Inagaki, K., & Hatano, G. (1993). Young children's understanding of the mind-body distinction. *Child Development, 64*, 1534–1549.

Inagaki, K., & Hatano, G. (1996). Young children's recognition of commonalities between animals and plants. *Child Development, 67*, 2823–2840.

Inagaki, K., & Hatano, G. (1991). Constrained person analogy in young children's biological inference. *Cognitive Development, 6*, 219–231.

Inagaki, K. (1990). The effects of raising animals on children's biological knowledge. *British Journal of Developmental Psychology, 8*, 119–129.

Johnson, S. C., & Carey, S. (1998). Knowledge enrichment and conceptual change in folkbiology: Evidence from Williams syndrome. *Cognitive Psychology, 37*(2), 156–200.

Johnson, S. C., & Solomon, G. E. A. (1997). Why dogs have puppies and cats have kittens: The role of birth in young children's understanding of biological origins. *Child Development, 68*(3), 404–419.

Kalish, C. W. (1996a). Causes and symptoms in preschoolers' conceptions of illness. *Child Development, 67*, 1647–1670.

Kalish, C. W. (1996b). Preschoolers' understanding of germs as invisible mechanisms. *Cognitive Development, 11*, 83–106.

Kalish, C. W. (1997). Preschoolers' understanding of mental and bodily reactions to contamination: What you don't see can hurt you, but cannot sadden you. *Developmental Psychology, 33*(1), 79–91.

Keil, F. C. (1989). *Concepts, kinds, and cognitive development.* Cambridge, MA: MIT Press.

Keil, F. C. (1994). The birth and nurturance of concepts by domains: The origins of concepts of living things. In L. Hirschfeld & S. Gelman (Eds.), *Domain specificity in cognition and culture* (pp. 234–254). New York: Cambridge University Press.

Keil, F. C., Levin, D. T., Richman, B. A., & Gutheil, G. (1999). Mechanism and explanation in the development of biological thought: The case of disease. In D.L. Medin & S. Atran (Eds.), *Folkbiology* (pp. 285–319). Cambridge, MA: MIT Press.

Kitcher, P. (1988). The child as parent of the scientist. *Mind and Language, 3*, 217–228.

Kuhn, T. S. (1983). Commensurability, comparability, and communicability. In P. Asquisth & T. Nickles (Eds.), *PSA 1982*. East Lansing, MI: Philosophy of Science Association.

Laurendeau, M., & Pinard, A. (1962). *Causal thinking in the child: A genetic and experimental approach*. New York: International Universities Press.

Lopez, A., Gutheil, G., Gelman, S. A., & Smith, E. E. (1992). The development of category-based induction. *Child Development, 63*, 1070–1090.

Medin, D. L., & Coley, J. D. (1998). Concepts and categorization. In J. Hochberg & J. E. Cutting (Eds.), *Handbook of perception and cognition: Perception and cognition at century's end* (2nd ed., pp. 403–440). San Diego: Academic Press.

Medin, D. L., & Ortony, A. (1989). Psychological essentialism. In S. Vosniadou & A. Ortony (Eds.), *Similarity and analogical reasoning* (pp. 179–195). Cambridge: Cambridge University Press.

Osherson, D. N., Smith, E. E., Wilkie, O., López, A., & Shafir, E. (1990). Category-based induction. *Psychological Review, 97*, 185–200.

Piaget, J. (1929). *The child's conception of the world*. London: Routledge and Kegan Paul.

Richards, D. D., & Siegler, R. S. (1984). The effects of task requirements on children's abilities to make life judgments. *Child Development, 55*, 1687–1696.

Richards, D. D. (1989). The relationship between the attributes of life and life judgments. *Human Development, 32*(2), 95–103.

Rosengren, K. S., Gelman, S. A., Kalish, C. W., & McCormick, M. (1991). As time goes by: Children's early understanding of growth in animals. *Child Development, 62*, 1302–1320.

Ross, N., Medin, D., Coley, J. D., & Atran, S. Cultural and experiential differences in the development of folk biological induction. Manuscript under review, *Child Development.*

Rozin, P., Fallon, A. E., & Augustini-Ziskind, M. L. (1985). The child's conception of food: The development of contamination sensitivity to "disgusting" substances. *Developmental Psychology, 21*, 1075–1079.

Solomon, G. E. A., & Cassimatis, N. L. (1999). On facts and conceptual systems: Young children's integration of their understanding of germs and contagion. *Developmental Psychology, 35*, 113–126.

Solomon, G. E. A. (1996). Race and naïve biology. Manuscript.

Solomon, G. E. A., Johnson, S. C., Zaitchik, D., & Carey, S. (1996). Like father, like son: Young children's understanding of how and why offspring resemble their parents. *Child Development, 67*(1), 151–171.

Springer, K., & Keil, F. C. (1989). On the development of biology specific beliefs: The case of inheritance. *Child Development, 60*(3), 637–648.

Springer, K. (1995). How a naïve theory is acquired by inference. *Child Development, 66*, 547–558.

Springer, K. (1996). Young children's understanding of a biological basis for parent-offspring relations. *Child Development, 67*, 2841–2856.

Suzuki, D., & Knudtson, P. (1992). *Wisdom of the elders: Sacred native stories of nature*. New York: Bantam Books.

Weissman, M. D., & Kalish, C. W. (1998). The inheritance of desired characteristics: Preschoolers' beliefs about the role of intention in biological inheritance. *Journal of Experimental Child Psychology, 73*, 245–265.

Wellman H. M., & Gelman, S. A. (1998). Knowledge acquisition in foundational domains. In D. Kuhn & R. S. Siegler (Eds.), *Handbook of child psychology, Vol. 2, Cognition, perception and language* (5th ed., pp. 523–573). New York: Wiley.

4

Children's Affiliations with Nature: Structure, Development, and the Problem of Environmental Generational Amnesia

Peter H. Kahn, Jr.

> The only time that I've seen dolphins in the Rio Tejo was by chance. The dolphins followed the boat. It was something that I've never forgotten. . . . It is one of those things that remain engraved in the memory. (Portuguese college student)

How do children reason about environmental problems? Are there universal features in children's environmental conceptions and values? How important is it that children and young adults—like the Portuguese student above who remembers having seen a dolphin in the Rio Tejo—experience natural wonders? Finally, what happens to children's environmental commitments and sensibilities when they grow up in environmentally degraded conditions?

In this chapter, I address these questions by drawing on the results of five studies my colleagues and I have conducted. In these studies, we interviewed children in diverse locations about their environmental moral conceptions and values. I also seek to explicate two ideas that frame my theoretical approach to investigating children's affiliations with nature—structure and development. Finally, I build on the structural-developmental framework and on my research findings to articulate what may be one of the most pressing and unrecognized problems of our age—the problem of environmental generational amnesia.

Structure and Development

When talking about a child's development, we often ask, "How did this child get to be this way?" And often we answer with one of two choices—either by nature or by nurture. But a third choice is possible—

that development arises not simply by nature or nurture (or some combination of both) but by the active mental constructions of children and the ways in which children organize and act on their knowledge and values.

Consider, for example, an infant who sees a small ball, reaches with one hand, and picks it up. Indeed, she can pick up a small ball with either hand and on many occasions does so. Now let us say that one day she encounters a balloon that she wants to pick up, but when she reaches out with one hand to grasp and lift it, she is unable to. She becomes disequilibrated. She has the interest and desire to pick up that balloon. Thus she struggles for a more adequate understanding. Maybe she tries repeatedly with the other hand, and that fails, too. At some point, she discovers a solution. She coordinates her two separate grasping schemes into a single consolidated scheme, and—in a remarkable developmental achievement—picks up the balloon using two hands.

Such a characterization of learning helps convey the tenor of *structural-developmental theory* (Damon, 1977; DeVries & Zan, 1994; Kohlberg, 1969; Piaget, 1983; Turiel, 1998). This theory is also sometimes known as *constructivist*, *social cognitive*, or *structural interactional.* Structural-developmental theory posits that through interaction with a physical and social environment children construct conceptual understandings and values. We can call these mental constructions *structures.* Structures develop. Moreover, through structural development early forms of knowledge do not disappear but are transformed into more comprehensive and adequate ways of understanding the world and of acting on it. Notice in the example above, for example, that the infant's earlier form of knowledge is not lost in development. She can still pick up a small object with either hand. But this knowledge is now hierarchically integrated into a larger conceptual organization.

I would like to flesh out these ideas about structure and development and show how they can be used to investigate children's affiliations with nature. To do so, I draw on five studies my colleagues and I have conducted, wherein we interviewed children about their environmental moral conceptions and values. In one study (the Prince William Sound Study) we interviewed children in grades two, five, and eight in Houston, Texas, about the oil spill that occurred in 1989 in Prince William Sound,

Alaska (Kahn, 1997a). In a second study (the Houston Child Study) we interviewed children in grades one, three, and five in an inner-city African American community in Houston, Texas (Kahn & Friedman, 1995). In a third study (the Houston Parent Study) we interviewed parents of the children in that same community (Kahn & Friedman, 1998). In a fourth study (the Amazonia Study) we interviewed (in Portuguese) children in grade five in an urban and a remote part of the Brazilian Amazon region (Howe, Kahn & Friedman, 1996). In a fifth study (the Lisboa Study) we interviewed (in Portuguese) children and young adults in grades five, eight, 11, and college in Lisbon, Portugal (Kahn & Lourenço, in press).

Before looking at some of the results, however, it should be noted that, methodologically, asking children questions that are identical to one's research question rarely leads to success. For example, the researcher who asks children "What is your conception of morality?" quickly finds that children have little to say. Rather, as illustrated in the moral-developmental research programs of Kohlberg (1984), Turiel (1983, 1998), and others, one first needs to demarcate the moral domain and then provide numerous moral stimuli that allow children ready access to moral concepts. So, too, with investigating children's affiliations with nature. Six overarching topics were pursued in the Lisboa Study:

- One series of questions focused on children's relationships to domestic animals ("Are pets important or not important to you?"), wild animals ("Are wild animals important or not important to you? What's the difference in your relationship to pets and wild animals?"), plants ("Are plants important or not important to you?"), parks ("Are the parks that exist around town important or not important to you?"), and environmental problems ("Do you know of any problems that affect the environment? If so, which ones? Do you talk about the problems with your friends or with your family? Do you do anything to protect the environment or to help solve some of the problems?").
- A second series of questions focused on whether children believed that throwing trash into their local river (the Rio Tejo) violated a moral obligation.
- A third series of questions focused on ways participants believed that throwing garbage into the Rio Tejo would harm fish, birds, the water,

the view of the landscape, and the people who lived beside the river (e.g., "Do you think throwing garbage in the Rio Tejo would affect the fish? How?") and whether children cared if such harm occurred (e.g., "Does it matter to you that the fish would be affected in this way?").

• A fourth series of questions focused on how children resolved potentially contradictory environmental judgments (e.g., "If driving a car causes pollution, and you said it is not all right to pollute, then is it all right or not all right to drive a car?").

• A fifth series of questions focused on what counts as "natural" activity (e.g., "If a fire in the forest is caused by lightning, would you say that the fire is natural?" and "If a fire in the forest is caused accidentally by a person, would you say that the fire is natural?").

• A sixth series of questions examined children's conception of harmony with nature ("Is it possible to live in harmony with nature and to cut down the trees in the forests? How?" and "For you, what does it mean to live in harmony with nature?").

Two overarching forms of reasoning emerged in all five studies for why children believed that nature should be valued—anthropocentric and biocentric. *Anthropocentric reasoning* refers to an appeal to how effects to the environment affect human beings. For example, consider the following adolescent's justification for why it is wrong to pollute the Rio Tejo:

> Look, . . . it is a very selfish theory. . . . From an economic point of view the [polluted] water would be captured and sent to a central plant where it would be treated. Who is paying for the process to clean the water? Isn't it us? So we are causing harm to ourselves.

In this response, the underlying reason that water pollution is wrong is that it harms human economic welfare. Other appeals to anthropocentric welfare included human physical welfare (air pollution is wrong "because the air is polluted, it is harder to breathe, and it can cause many more diseases") and human psychological welfare (domestic animals are important because "if they belong to a child, they can contribute to his or her development"). Besides welfare, other anthropocentric justifications included appeals to personal human interests ("because if the Rio Tejo were clean, we could swim in it"), human-centered justice considerations ("nobody has the right to make [the water] dirty, it belongs to the public"), and aesthetics ("because dirty water is unpleasant, there is no comparison to see a river with clean water, to see the fish swimming, to see the pebbles, and to see that brown, grayish, thick disgusting water").

Biocentric reasoning refers to an appeal that the natural environment has moral standing that is at least partly independent of its value as a human commodity. For example, one form of biocentric reasoning focuses on the intrinsic value of nature and establishes that value by means of a *teleos*—a proper functioning or preordained endpoint: wild animals "are important because if someone created them, it is because they have some kind of role." In this response, the adolescent reasons that animals are important based on their preordained place in the world. Other biocentric justifications focus on an appeal that nature has rights, deserves respect or fair treatment, or merits freedom (wild animals are important "because I think that all animals have the right to their life"). I will provide further examples of biocentric reasoning shortly.

In our published scientific papers on how children value nature, my colleagues and I have analyzed and reported on the justification data quantitatively. We have, for example, reported percentages of each justification for each specific question. (For more details about all the studies, see Kahn, 1999.) In general, we found a comparatively large use of anthropocentric reasoning (roughly 95 percent) and a small use of biocentric reasoning (roughly 5 percent). We also found this pattern to occur in the Amazonia Study, which included a population of children who lived in a small village along the Rio Negro that is inaccessible except by boat. This finding was surprising because it could reasonably be expected that children who live intimately with nature would have a greater biocentric affiliation with the land and animals. Instead, only in the Lisboa Study—which included an adolescent and college-age population—did we find that certain questions pulled more biocentric responses than anthropocentric responses. Specifically, in response to the question of whether wild animals are important or not important, 73 percent of the justifications were biocentric. In response to the question of why people should care if birds are harmed by water pollution, 34 percent were biocentric. While the first question was not asked in our other studies (thus preventing a direct comparison), my interpretation of these data includes the following developmental explanation.

Recall from the infant scenario above that with the coordination of two independent grasping schemes an infant does not lose access to each scheme independently: although she has learned to use two hands to grasp a balloon, she still can grasp a ball with a single hand. It is possible that through development a child's unelaborated concerns give way to both anthropocentric considerations of human welfare and early biocentric considerations that focus on the intrinsic value of nature (animals are important "because they are living beings"). The mental organization of each group of considerations initially can be considered a structure. But I propose that a more advanced biocentric structure comes to encompass these two earlier structures (which we can now think of as partial structures) and their subsequent coordinations. In other words, children with comparatively advanced biocentric reasoning coordinate anthropocentric and early biocentric considerations, while being able to draw—in different contexts—on each partial structure by itself.

A clear example of these coordinations can be seen in what Lourenço and I have called *isomorphic biocentric reasoning*. Here an appeal is based on recognizing a correspondence between humans and animals, either by means of direct or conditional considerations. In a *direct isomorphism*, humans and nature are viewed as essentially similar, and sometimes the relevant properties are specified; accordingly, an appeal is made that nature thereby deserves the same moral consideration as humans. For example, Jill, a participant from the Prince William Sound Study, said, "I think fish and animals have a right to live just like we do, and it's not fair to have killed them this way." In this response, Jill establishes a symmetrical correspondence between humans and nature (the right to life), which leads to a judgment of unfairness. In turn, a *conditional isomorphism* establishes a direct isomorphism by means of an if-then conditional judgment. For example, a participant in the Lisboa Study said, "If we don't like to live surrounded by trash, [then the fish] don't like it also."

Developmentally, however, isomorphic reasoning does not appear to represent an endpoint. Imagine, for example, if we had pressed Jill (above) with moral counterclaims to her statement that animals have a right to live as humans do. Let us say we had asked, "What if a person had a health problem that improved when he or she ate fish? Would it then be all right to eat fish?" Jill could say, "No, fish have the same right to live as we have." We could have also asked, "What if a person is on a deserted island, and this is her only way to live?" Again, Jill could say, "No, fish have the same right to live just as we have." Indeed, similar conversations have occurred in our interviews, and elsewhere (Kahn, 1999, chap. 6) I have suggested that when such counterclaims gain purchase in a child's psyche, the claims initiate the disequilibration that leads to development.

In turn, transmorphic reasoning takes an isomorphism and then extends it through either compensatory or hypothetical considerations. In a *compensatory transmorphism*, similarities are coordinated with differences. For example, a participant in the Lisboa Study said, "[Wild animals are important] because they breathe like we do, and sometimes we think that because they are animals, they are not like us, that they don't do certain things. Then we end up seeing that they do." This participant understands that animals are in certain respects different from humans ("they don't do certain things") but also similar ("they breathe like we do") and that such differences do not void a mapping of similar value considerations from humans to nature. In a *hypothetical transmorphism*, principled reasoning includes impartiality and generalizability as organizing features of the environmental moral judgment. For example, a participant in the Prince William Sound Study said, "You put yourself in the animal's position, and you wouldn't like that. And so if you just kind of trade places and think about it, and everyone would think it wasn't right."

Though I do not yet have enough fine-grained data that would allow for a developmental analysis, my sense is that transmorphic reasoning hierarchically integrates isomorphic reasoning. In other words, in development the ability to conceptualize a relationship between humans and animals in symmetrical terms is not lost but integrated into a more comprehensive structure that can account for asymmetrical characteristics.

Taken more broadly, this account of the coordinations of anthropocentric considerations and early biocentric considerations helps provide a specific answer to the question of which comes first in children's development—a moral relationship with animals or with people. I think the answer is neither one but both, dialectically (cf. Myers, 1998). In other words, children's moral relationships with other animals help establish their moral relationships with people and vice-versa.

One more example of a possible hierarchical integration will prove interesting, particularly because it highlights a different aspect of children's affiliations with nature. Across several of the studies, we asked children to describe what it means to them to live in harmony with nature and to provide us with some examples. Five overarching categories of conceptions of harmony emerged from the results—physical, sensorial, experiential, relational, and compositional.

A *physical* conception is based on doing something to nature, for nature, or with nature. It includes negative acts ("Harmony with nature is not to destroy trees, not to destroy nature"), positive acts ("Harmony means to protect the animals and the plants"), and activity ("When a person is living in harmony with nature, he goes to the countryside and has a picnic"). A *sensorial* conception is based on apprehending nature directly with the senses ("Harmony means seeing everything blooming, not seeing people cutting trees down, smelling nature's environment"). An *experiential* conception is based on experiencing a particular state of mind or feeling ("Harmony means feeling comfortable with yourself in that moment and in that place"). A *relational* conception is based on a relationship between humans and nature ("[Harmony means] talking with the trees. . . . Sometimes I talk to them as if they were people, like this"). Finally, a *compositional* conception is based on being in balance with nature. It includes a focus on anthropocentric compositions ("We can live in harmony with nature without having to destroy more than we are allowed; nature has X resources to give us, and if we take them all at once, we leave nothing to grow") and on biocentric compositions ("To live in harmony, it is the balance. We trade with nature in a way that none of the parts suffer any harm").

To be clear, by "a compositional conception" of harmony, I mean something like a musical or artistic composition whose parts support the integrity, beauty, balance, and proportion of the whole. Given this definition, which embeds within it physical, sensorial, experiential, and relational concepts, compositional reasoning may hierarchically integrate at least some of the earlier categories. This proposition is strengthened by the developmental findings in the Lisboa Study (which included the oldest age groups)—namely, that the use of compositional reasoning increased with age: fifth grade (3 percent), eighth grade (31 percent), eleventh grade (52 percent), and college (71 percent).

Cross-Cultural Comparisions

Across three of our studies—the Houston Child Study, the Amazonia Study, and the Lisboa Study—we asked children some of the same questions. In this way, we were able to perform direct cross-cultural comparisons. Results showed that children across these three studies often demonstrated remarkably similar environmental moral values and knowledge. For example, the large majority of children in all three locations believed that animals and plants were important in their lives; were aware of environmental problems that affected themselves or their community; believed that throwing garbage into their local waterway harmed birds, the view, and the people who lived along the river; cared that such harm might occur; and, based on the criterion judgments of prescriptivity, rule contingency, and generalizability, believed it was a violation of a moral obligation to throw trash into their local waterway.

Moreover, time and again, as my colleagues and I read through the interviews with children, we felt the structural similarities—the similar organization of children's reasoning—across diverse locations. To provide a sense of what we have been looking at, consider the following sets of justifications:

1A. "Because some people that don't have homes, they go and drink out of the rivers and stuff, and they could die because they get all of that dirt and stuff inside of their bodies." (Houston Child Study)

1B. "Because it causes pollution, that is dangerous for us. Because now we have cholera, a very dangerous disease, and there are others attacking us, like the malaria." (Amazonia Study)

1C. "Because it would harm the health of everybody using that water either to drink or to bathe, anything at all." (Lisboa Study)

All three of the above children reason that it is wrong to throw garbage into their local waterway because people might drink from polluted water and get sick ("they could die"; "now we have cholera, a very dangerous disease"; "it would harm the health of everybody").

2A. "Because water is what nature made; nature didn't make water to be purple and stuff like that, just one color. When you're dealing with what nature made, you need not destroy it." (Houston Child Study)

2B. "Because the river was not made to have trash thrown in it, because the river belongs to nature." (Amazonia Study)

2C. "Because the river was not created [for people] to throw trash into it. It is a natural means, another natural means that should not be destroyed." (Lisboa Study)

All three of the above children base their environmental judgments on the idea that nature has its own purpose ("nature didn't make water to be purple and stuff"; "the river was not made to have trash thrown in it"; "the river was not created to throw trash into it").

3A. "Some people don't like to be dirty. And when they throw trash on the animals, they probably don't like it. So why should the water be dirty and they don't want to be dirty?" (Houston Child Study)

3B. "Because animals have to have their chance. They also must have to live. We should not mistreat them because if it happens to us, we don't like it." (Amazonia Study)

3C. "They [plants] are important, as the animals are important, because they are living beings and live like us." (Lisboa Study)

All three of the above children establish isomorphic relationships. They judge the mistreatment of animals or plants to be wrong by considering whether humans would like to be treated in a similar way ("some people don't like to be dirty . . . [so the animals] probably don't like it"; "because if it happens to us, we don't like it"; "they are living beings and live like us").

4A. "Fish don't have the same things we have. But they do the same things. They don't have noses, but they have scales to breathe, and they have mouths like we have mouths. And they have eyes like we have eyes." (Houston Child Study)

4B. "Even if the animals are not human beings, for them they are the same as we are. They think like we do." (Amazonia Study)

4C. "Because they breathe like we do, and sometimes we think that because they are animals, they are not like us, that they don't do certain things. Then we end up seeing that they do." (Lisboa Study)

All three of the above children establish transmorphic relationships. They recognize that while animals are not identical to human beings ("fish don't have the same things we have"; "animals are not human beings"; "they are not like us") that both animals and people have significant functional equivalences ("[fish] don't have noses, but they have scales to breath"; "[animals] think like we do"; "[animals] breath like we do").

In short, our results support the proposition that across cultures children's affiliations with nature are often similarly structured.

It is important to recognize that humans have both positive and negative experiences with nature. We investigated negative experiences most directly in terms of water pollution, air pollution, and garbage. As noted above, we found that children, whether living in an economically impoverished urban African American community (the Houston Child Study) or a relatively pristine rain forest (one of the populations in the Amazonia Study), often used anthropocentric welfare justifications to appeal to the human need for clean water to drink and clean air to breath. Such reasoning was also central to the adults we interviewed in an African American community in Houston. As one adult said in the Houston Parent Study:

> [The air] stinks 'cause I laid up in the bed the other night, kept smelling something. Knew it wasn't in my house 'cause I try to keep everything clean. Went to the window, and it almost knocked me out. The scent was coming from outdoors into the inside, and I didn't know where it was coming from. . . . Now, who'd want to walk around smelling that all the time?

Thus it is possible that pollution offers one of the most direct negative experiences that people commonly have with nature and that people everywhere who recognize such pollution can be expected to object to it. I return to this idea in the next section.

From our data, it would also appear the humans affiliate with positive aspects of nature. For example, across the Houston Child Study, the Amazonia Study, and the Lisboa Study the large majority of children said that animals and plants played an important part in their lives and that they cared about the well-being of birds and landscape aesthetics. Similar positive affiliations emerged in the Houston Parent Study. For illustrative purposes, consider the similarity of reasoning between a Portuguese college student and an African American parent in the inner city of Houston:

> I live in the country, and I find that living in the city is very difficult. It causes stress. For instance, we live on this street full of trees. Anytime that I leave home in the morning, I feel invigorated seeing the trees and their shade. I can breathe. I can hear the birds. Now, if I lived on a street close to Avenida da Republica, I would feel stressed seeing that amount of cars, very few trees. (Lisboa Study)

> Yesterday, as my son and I were walking to the store and we were walking down Alabama [Street], and for some reason, I think they're getting ready to widen the street. And it's a section of Alabama that I thought was so beautiful because of the trees, and they've cut down all the trees. And you know it hurts me every time I walk that way, and I hadn't realized that my son had paid attention to it, too. (Houston Parent Study)

Both participants express appreciation for trees, especially in the context of living in a congested city.

If it is true that many forms of environmental reasoning—and more broadly, negative and positive affiliations with nature—cut across cultures, then why is this so? One answer draws on sociobiological theory and looks like this. Imagine having lived on the savannas of East Africa, as human beings did for nearly 2 million years. If you wanted to survive, it would be good to be scared of snakes that could kill you, and it would be good to be attracted to clean bodies of water so that you could drink and to plants and animals so that you could eat. In other words, in the standard sociobiological account (Wilson, 1975, 1984, 1998), genes that have led to certain negative affiliations with nature (disliking polluted water and poisonous snakes) and positive affiliations with nature (enjoying trees and the beauty of flowers) have enhanced survival and have tended to reproduce themselves since they have been in bodies that have procreated more rather than less. Thus, these genes, correlative affiliations, and resulting behaviors have grown more frequent.

In my view, the sociobiological answer is right, up to a point. We are biological beings with an evolutionary history, and any account of children's affiliation with nature needs to build from this perspective. But as

I have argued elsewhere (Kahn, 1997b, 1999), biology, genes, and genetic fitness do not go far enough, pragmatically and theoretically. Pragmatically, we as a species can make bad choices and become extinct. Theoretically, we need to account for concepts of intentionality, free will, meaning, and the possibility for individuals to shape—from an ethical stance—cultural practices.

Another answer to the question of why we found so much similarity in environmental moral reasoning across cultures draws on structural-developmental theory. Recall that structural-developmental theory is an interactional theory: through interaction with a physical and social world children construct knowledge and values. Thus, it seems plausible that certain features of the natural environment are pervasive enough across diverse contexts to allow for the development of similar constructions. Even, for example, in the inner city of Houston—where human violence and drugs were an everyday part of children's experience—children interacted with vibrant parts of the natural world. As one participant in the Houston Parent Study said:

> My kindergarten daughter, she might see something that looks injured, and um she saw a worm. She doesn't pick up these black ones or brown ones because they sting. So this one was a yellow one, and she said he was hungry. So she picked him up and took him over to a leaf and put him on it. You know, they do those type things.

Bugs, pets, plants, trees, wind, rain, soil, sunshine: such manifestations of nature occur not only in the Brazilian Amazon but in our cities.

Environmental Generational Amnesia

I have suggested that similar manifestations of nature occur across diverse locations and that such similarities help explain children's similar environmental moral constructions. But I want to be careful here, for this proposition might seem to imply that we can continue to degrade the environment with impunity. After all, if there were few differences in environmental reasoning and values between children growing up in an economically impoverished urban community in Houston and in a relatively pristine village in the Amazon rain forest, then—at least in terms of nature's impact on children's development—do we really have to worry about nature's destruction?

I first began to understand why the answer is yes when looking at several findings from the Houston Child Study. Houston is one of the more environmentally polluted cities in the United States. Local oil refineries contribute not only to the city's air pollution but also to distinct oil smells on many days. Bayous can be thought of more as sewage transportation channels than freshwater rivers. Within the community where we conducted the Houston Child Study, garbage was commonly found alongside the bayou and on the streets and sidewalks. With that said, colleagues and I systematically investigated whether children who understood in general about the idea of air pollution, water pollution, and garbage also understood that they directly encountered such pollution in Houston. The findings showed a consistent statistically supported pattern. About two-thirds of the children understood in general about these three environmental problems. However, contrary to our expectations, only one-third of the children believed that these environmental problems affected them directly.

How could children who know about pollution in general and live in a polluted city be unaware of their own city's pollution? One answer is that to understand the idea of pollution one needs to compare existing polluted states to those that are less polluted. In other words, if one's only experience is with a certain amount of pollution, then that amount becomes not pollution but the norm against which more (or less) polluted states can be measured at a later time. The crux here is that like the children in Houston, I think we all take the natural environment we encounter during childhood as the norm against which we measure environmental degradation later in our lives. With each ensuing generation, the amount of environmental degradation increases, but each generation in its youth takes that degraded condition as the nondegraded condition—as the normal experience. I have called this psychological phenomenon *environmental generational amnesia* (Kahn, 1997b, 1999; Kahn & Friedman, 1995).

I said I would come back to the idea that pollution offers one of the most direct negative experiences that people commonly have with nature and that people everywhere who recognize such pollution can be expected to object to it. Now we can see that this idea is not as straightforward as it might appear. For one thing, children might not recognize such pollution. For another thing, people's objections across generations may not keep pace with worsening environmental conditions.

Environmental generational amnesia offers a different perspective on what many observers of the global human condition view as environmental complacency. For example, after his visit to some of the most polluted cities in China, Hertsgaard (1998, p. 158) wrote that "while there were plenty of things the Chinese masses might not like about their existence, by far their biggest complaint was being miserably poor, and they would put up with a great deal of aesthetic or environmental unpleasantness to escape poverty." Along similar lines, Huber (quoted in The Greening of Affluent America, 2000) argues that people become environmentally oriented "when they feel personally secure, when their own appetites have been satisfied, when they do not fear for the future, or for their own survival, or their children's. . . . It is the rich who can be green because they no longer have to choose between their own survival and nature's." In other words, with at least an implicit nod to Maslow's hierarchy of values, a common argument is that first people need to feed their bellies and only then can they become concerned with higher-order values, such as environmental degradation.

But I do not think environmental complacency can be adequately understood in such terms. Rather, consider what my colleagues and I saw emerge in the Houston Parent Study. We asked parents, on a scale of 1 to 10 (with 1 the least important and 10 the most important), to rank the importance of drug education for their children. Results showed a mean rank of 8.5 (standard deviation 3.3). On the same scale, we asked parents to rank the importance of environmental science education for their children. Results showed a mean rank of 8.7 (SD 2.4). Statistical tests showed no difference. Of parents who equated the importance of drug education and environmental science education, their reasoning often focused on the physical ramifications of both problems:

> With the drugs, we're nothing. Without the environment, we're nothing. And drugs is something I see every day. There are dealers across the street from me. So I see this every day, and it's just killing us. I mean, it really is killing us, and with the drugs, we're not going to have any youth. . . . With the drugs, you're not going to have a future, and without any environment we're not going to have a future.

Well, let's put it like this here. If you don't take care of one [drugs], it's going to kill you. If you don't take care of the other [the environment], it's going to kill you.

These findings are of a piece with the environmental justice literature (Bullard, 1990; Faber, 1998b; Mohai & Bryant, 1992) that documents the ongoing struggle by poor people and people of color to protect their "land, water, air, and community health [from] corporate polluters and indifferent governmental agencies" (Faber, 1998a, p. 1). Thus while poverty surely affects certain aspects of people's environmental behavior (such as whether they pay the higher prices of organically grown foods), I do not think environmental complacency is caused simply by poverty. Rather, because of environmental generational amnesia, I think we all have difficulty understanding in a direct, experiential way that we have environmental problems of any magnitude.

Historically, this explanation seems to fit. For example, many centuries ago the forests in the highlands of Scotland flourished. According to Hand (1997, p. 11), these forests were

grand as any on earth. Elm, ash, alder, and oak shaded the low-lying coastal plains and inland valleys; aspen, hazel, birch, rowan, and willow covered the hills; and beautiful, redbark Scots pine clung to the glacial moraines and steep granite slopes. The Romans called it the Forest of Caledonia, "the woods on heights," and it clung to Scottish soil for millennia.

However, at the start of the sixteenth century, with the coming of the English and the industrial revolution, the forests came under siege, and by the 1700s they had been virtually eliminated (ibid., p. 12):

Stone houses and coal fires replaced those of wood. Soils, exposed to harsh winds and rain, washed into streams and rivers, leaching fertility, destroying fisheries. Erosion cut, in many places, to bedrock. Woodland species—bear, reindeer, elk, moose, beaver, wild boar, wild ox, wolf (the last killed in 1743), crane, bittern, great auk, goshawk, kite, and seaeagle—vanished. . . . By 1773, when Dr. Samuel Johnson toured the highlands, with James Boswell, the landscape was, in Johnson's words, a "wide extent of hopeless sterility." He remarked that one was as likely to see trees in Scotland as horses in Venice."

Today the highlands of Scotland are one of the most deforested lands in the world. Perhaps equally disturbing, the Scots of today, according to Hand, have virtually no conception of a forest, of its ecological vastness and beauty. Hand presented these ideas in an essay titled "the forest of forgetting." It is a forgetting that crosses generations.

Environmental generational amnesia also appears to affect even the most environmentally vocal. Take a guess, for example, when the following magazine editorial was written:

> This [society] is born of an emergency in conservation which admits of no delay. It consists of persons distressed by the exceedingly swift passing of wilderness in a country which recently abounded in the richest and noblest of wilderness forms, the primitive, and who purpose to do all they can to safeguard what is left of it.

In the last decade we have indeed witnessed the swift passing of wilderness in the United States, and environmentalists often speak of this problem as one that "admits of no delay." The above passage was written, however, in 1935 as the opening to the first issue of *The Wilderness Society* (1993, p. 6). Thus environmental problems can be described as equally serious across generations even while the problems worsen.

If it is difficult for us to construct accurate understandings about our negative experiences with nature—when such experiences can have us choking on our air and drinking bottled water—then it is all the more difficult for people to construct accurate understandings about their loss of positive affiliations with nature. Meloy (1997, pp. 4–5), writes, for example, that in 1929 her mother, then a child,

> bellied up to the edge of a sheer cliff on a 14,495-foot Sierra peak and, while someone held her feet, stared down into empty blue-white space. Local newspapers reported her as the first child to climb Mt. Whitney. "On that three-week trip we saw one other pack train from a distance," [her mother] recalled, "and we said the mountains were getting crowded." . . . [Now] thirty million people live within a day's drive of Sequoia and Kings Canyon parks. Space atop Mt. Whitney is rationed: you need a reservation to climb it from the east.

Yet people today still speak of such outings in Kings Canyon as "wilderness" outings, and a packed freeway in the middle of Los Angeles can be referred to as "noncongested" as long as the cars are moving along in a timely fashion. Apparently, environmental generational amnesia also leads us to construct distorted meanings for environmental concepts.

As we continue to degrade nature, we will adapt to its loss, as we have already, no doubt. But the adaptation comes with physical and psychological costs.

Consider this analogy. Imagine that your favorite food item is the only source of an essential nutrient and that without it everyone suffers from low-grade asthma and increased stress. Now imagine a generation of people who grow up in a world where this food item does not exist. In such a world, it would seem likely that people would not feel deprived by the absence of this tasty food (it was never in their minds to begin with) and that they would accept low-grade asthma and increased stress as the normal human condition.

Nature is like that food. A wide variety of literature, which has come under the rubric of *biophilia*, shows that direct positive affiliations with nature have beneficial effects for people's physical, cognitive, and emotional well-being (Kellert & Wilson, 1993; Wilson, 1984). Findings from over 100 studies, for example, have shown that stress reduction is one of the key perceived benefits of recreating in a wilderness area (Ulrich, 1993). Other studies have shown greater stress recovery in response to natural than urban settings (Ulrich, Simons, Losito, Fiorito, Miles & Zelson, 1991). Other studies conducted in prisons, dental offices, and hospitals point to similar effects. For example, Moore (1982; cited in Ulrich, 1993) found that prison inmates whose cells looked out onto nearby farmlands and forests needed fewer health care services than inmates whose cells looked out onto the prison yard. In short, the research literature shows that people who affiliate positively with nature tend to be happier, more relaxed, more productive, more satisfied with their homes and jobs, and healthier. In Kaplan and Kaplan's (1989, p. 198) reading of this literature, they write that as "psychologists we have heard but little about gardens, about foliage, about forests and farmland. . . . Perhaps this resource for enhancing health, happiness, and wholeness has been neglected long enough."

Solving the Problem of Environmental Generational Amnesia

How can we solve the problem of environmental generational amnesia? There is no easy answer. But one important thing to understand is that this problem has its genesis in childhood. And therein we must look for solutions.

The structural-developmental (constructivist) approach to education offers a starting point. Recall that this theory posits that children are not passive beings who are merely programmed genetically or molded societally but that through interaction with their environment, children construct knowledge and values. Thus constructivist education allows children to explore, interact, recognize problems, attempt solutions, make mistakes, and generate more adequate solutions. Moreover, constructivist education follows Kohlberg and Mayer's (1972) landmark dictum that development should be a central aim of education. This dictum helps speak to the importance of developmental research for discovering the pathways along which we can help guide our children.

In this light, structural-developmental research on children's affiliations with nature can be used proactively. I have characterized, for example, various forms of children's anthropocentric reasoning, including personal interests, physical welfare, psychological welfare, justice, and aesthetics. Developmentally speaking, these forms of reasoning are not wrong but incomplete; and I have suggested that when adolescents recognize certain limitations in their way of understanding their relationship with nature, biocentric conceptions can emerge through the hierarchical integration of these anthropocentric structures. Similarly, I suggested that compositional conceptions of living in harmony with nature integrated or at least emerged out of earlier conceptions—physical, sensorial, experiential, and relational.

Elsewhere I have offered a constructivist account of environmental education (Kahn, 1999, chap. 12; cf. Wals, 1994). But even with constructivist environmental education in place, the problem of environmental generational amnesia will persist. The reason is that by definition this problem arises because of an increasingly impoverished natural environment that limits the richness and diversity of a child's interaction with the natural world. Accordingly, one further response is to engage in dialogue with children about what has been lost and to use such dialogue to help shape the future. In this regard, consider the experiences of two college-age participants from the Lisboa Study:

> I heard that some time ago, when there was none of that pollution, the river was, according to what I heard, was pretty, there were dolphins and all swimming in it. I think it should have been pretty to see. Anyone would like to see it.
>
> I remember, for instance, a person who still talks about the time when he used to swim in the Rio Tejo and that he misses that a lot. And I, just eighteen years old, find it difficult to believe that this was possible. However, that was the main source of enjoyment of that person.

Granted, such dialogues can fall prey to adult monologues that romanticize the past and gripe about the present and future ("let me tell you how things were so much better when I was young"). But when such dialogues form part of engaged conversations, they have their place. They provide a means for children to gain information (otherwise unavailable in a direct experiential way) from which they can construct more veridical understandings of the natural world.

Along similar lines, teachers can use historical diaries and historical novels to convey a sense of the landscape of years past, and writing assignments can involve students in the comparative endeavor: "If you were the person in that historical novel and you saw the land today, what would you see, and what might you say?" Or students can work together to recover a piece of land nearby their school, bringing back native plants and biological heterogeneity. Nearby parks can be redesigned not as domesticated areas of extended lawn and play structures but as meadows, wetlands, forest, and creeks.

Equally important, we need to help children experience more pristine nature. This idea is captured in the thoughts of a child from the Lisboa Study:

> My grandmother lives in the north and I go there. And there are many rivers that still aren't polluted. And I think that, I go up there, and then I come back. I see up there a river that is not polluted. I feel the water running. I come back down here, I see trash. I think that there is such a difference. And I would like that the Rio Tejo—because I live in Lisbon, I was born in Lisbon—would like that the river in my hometown were not so polluted.

Of course, for such experiences to occur we need a more pristine nature for children to experience. Seen in this way, it becomes crucial to preserve pristine areas in settings both urban (parks and open areas) and rural (such as the Amazon rain forests). Such areas help provide the baseline of ecological health from which children (and societies at large) can construct notions of ecological disease.

Conclusion

Children construct rich and varied conceptions and values of the natural world, and they do so even in economically harsh urban settings. But as we degrade the environment, often for material gain, we are destroying the wellsprings of our children's psychological constructions.

This destructive process can be viewed clearly in the problem of environmental generational amnesia. To restate the basic idea: People take the natural environment they encounter during childhood as the norm against which they measure environmental degradation later in their life. With each ensuing generation, the amount of environmental degradation increases, but each generation takes that degraded condition as the nondegraded condition, as the normal experience. The upside of environmental generational amnesia is that each generation starts afresh, unencumbered mentally by the environmental misdeeds of previous generations. But the downside is enormous. As we lose daily, intimate positive affiliations with nature and accept negative experiences (such as pollution) as the norm, we suffer physically and psychologically and hardly know it.

What knowledge we have of nature often comes later in life and is hard won. Many of us as adults have found that our favorite outdoor place from our younger years has been lost. Perhaps a favorite tree has been cut down, or a favorite meadow paved. Perhaps our entire valley has become an epicenter of urban sprawl. Such experiences provide us with a basis for comparison and perhaps the impetus for environmental activism. But since each generation experiences only incremental harm, based on a comparison to a not too distant past, even our hard-won knowledge is incomplete, and so our sense of urgency often remains muted.

Since the problem of environmental generational amnesia has its genesis in childhood, I suggest that childhood is a good place to start solving the problem. We need to engage children in constructivist environmental education to maximize their exploration of and interaction with the nature that still exists within their purview—bugs, pets, plants, trees, wind, rain, soil, sunshine. We need to recognize that children's earlier forms of environmental reasoning are not usually wrong but incomplete and are capable of being transformed into more adequate forms of knowledge. We also need to recognize that children construct knowledge and values not only through interaction with a physical world (with nature) but through interaction with a social world and with social discourse.

Finally, the problem of environmental generational amnesia sets into motion a new and important argument for the preservation of the natural world. We need to design our cities with nature in mind, in view, and within grasp. We need open areas near cities, open ridge tops, public access to coastline, and city parks. We need to preserve pristine areas, wildlife areas, and wilderness areas—vast tracts of land as well as small tracts. With over 6 billion people on this planet, we are consuming land at an astonishing rate. We must recognize our need for a more pristine and at times wild nature so that adults and children alike can experience it, construct concepts of ecological health, and be nourished by it in body and mind.

Note

I thank Orlando Lourenço for his comments on an earlier version of this chapter and for his collaboration in the Lisboa Study.

References

Bullard, R. D. (1990). *Dumping in Dixie: Race, class, and environmental quality.* Boulder, CO: Westview Press.

Damon, W. (1977). *The social world of the child.* San Francisco: Jossey-Bass.

DeVries, R., & Zan, B. (1994). *Moral classrooms, moral children: Creating a constructivist atmosphere in early education.* New York: Teachers College, Columbia University.

Faber, D. (1998a). Introduction. In D. Faber (Ed.), *The struggle for ecological democracy: Environmental justice movements in the United States* (pp. 1–26). New York: Guilford Press.

Faber, D. (Ed.). (1998b). *The struggle for ecological democracy: Environmental justice movements in the United States.* New York: Guilford Press.

The greening of affluent America. Available at <http://www.hooverdigest.org/011/huber.html>.

Hand, G. (1997). The forest of forgetting. *Northern Lights, 13*(1), 11–13.

Hertsgaard, M. (1998). *Earth odyssey: Around the world in search of our environmental future.* New York: Broadway Books.

Howe, D., Kahn, P. H., Jr., & Friedman, B. (1996). Along the Rio Negro: Brazilian children's environmental views and values. *Developmental Psychology, 32*, 979–987.

Huber, P. (2000). *Hard green: Saving the environment from the environmentalists*. New York: Basic Books.

Kahn, P. H., Jr. (1997a). Children's moral and ecological reasoning about the Prince William Sound oil spill. *Developmental Psychology, 33*, 1091–1096.

Kahn, P. H., Jr. (1997b). Developmental psychology and the biophilia hypothesis: Children's affiliation with nature. *Developmental Review, 17*, 1–61.

Kahn, P. H., Jr. (1999). *The human relationship with nature: Development and culture*. Cambridge, MA: MIT Press.

Kahn, P. H., Jr., & Friedman, B. (1995). Environmental views and values of children in an inner-city black community. *Child Development, 66*, 1403–1417.

Kahn, P. H., Jr., & Friedman, B. (1998). On nature and environmental education: Black parents speak from the inner city. *Environmental Education Research, 4*, 25–39.

Kahn, P. H., Jr., & Lourenço, O. (in press). Air, water, fire, and earth: A developmental study in Portugal of environmental moral reasoning. *Environment and Behavior*.

Kaplan, R., & Kaplan, S. (1989). *The experience of nature: A psychological perspective*. Cambridge: Cambridge University Press.

Kellert, S. R., & Wilson, E. O. (Eds.). (1993). *The biophilia hypothesis*. Washington, DC: Island Press.

Kohlberg, L. (1969). Stage and sequence: The cognitive-developmental approach to socialization. In D. A. Goslin (Ed.), *Handbook of socialization theory and research* (pp. 347–480). New York: Rand McNally.

Kohlberg, L. (1984). *Essays in moral development*, Vol. 2, *The psychology of moral development*. San Francisco: Harper & Row.

Kohlberg, L., & Mayer, R. (1972). Development as the aim of education. *Harvard Educational Review, 42*, 449–496.

Meloy, E. (1997). Waiting its occasions. *Northern Lights, 13*(1), 4–6.

Mohai, P., & Bryant, B. (1992). Environmental racism: Reviewing the evidence. In B. Bryant & P. Mohai (Eds.), *Race and the incidence of environmental hazards: A time for discourse* (pp. 163–176). Boulder, CO: Westview Press.

Moore, E. O. (1982). A prison environments' effect on health care service demands. *Journal of Environmental Systems, 11*, 17–34.

Myers, G. (1998). *Children and animals: Social development and our connections to other species*. Boulder, CO: Westview Press.

Piaget, J. (1983). Piaget's theory. In W. Kessen (Ed.), *Handbook of child psychology*, Vol. 1, *History, theory, and methods* (4th ed., pp. 103–128). New York: Wiley.

Turiel, E. (1983). *The development of social knowledge*. Cambridge: Cambridge University Press.

Turiel, E. (1998). Moral development. In N. Eisenberg (Ed.), *Handbook of child psychology*, Vol. 3, *Social, emotional, and personality development* (5th ed., pp. 863–932). New York: Wiley.

Ulrich, R. S. (1993). Biophilia, biophobia, and natural landscapes. In S. R. Kellert & E. O. Wilson (Eds.), *The biophilia hypothesis* (pp. 73–137). Washington, DC: Island Press.

Ulrich, R. S., Simons, R. F., Losito, B. D., Fiorito, E., Miles, M. A., & Zelson, M. (1991). Stress recovery during exposure to natural and urban environments. *Journal of Environmental Psychology, 11*, 201–230.

Wals, A. E. J. (1994). *Pollution stinks!* De Lier, the Netherlands: Academic Book Center.

Wilderness Society. (1993). *56*(200), 6.

Wilson, E. O. (1975). *Sociobiology: The new synthesis*. Cambridge, MA: Harvard University Press.

Wilson, E. O. (1984). *Biophilia*. Cambridge, MA: Harvard University Press.

Wilson, E. O. (1998). *Consilience*. New York: Knopf.

5

Experiencing Nature: Affective, Cognitive, and Evaluative Development in Children

Stephen R. Kellert

This chapter is a largely theoretical examination of the role of experience and contact with nature in affective, cognitive, and evaluative (values-related) development among children during primarily middle childhood and early adolescence. Empirical and theoretical evidence is marshaled to support this conceptual framework, but a paucity of systematic and rigorous research suggests caution in accepting the conclusions and the need for future scientific study to test these concepts. Three kinds of experience of nature are distinguished in assessing possible developmental impacts on children—direct, indirect, and symbolic or vicarious experience. Additionally, the concept of *biophilia* (Kellert, 1997; Kellert & Wilson, 1993; Wilson, 1984) and a related typology of weak inherent tendencies to value nature are used to elucidate the role of childhood experience of nature in personality formation and character development. This chapter concludes by examining the possible developmental impacts of apparent declines in modern society of direct experience among children of abundant and healthy natural systems and the likely increase in indirect and vicarious contacts with the natural world.

This examination of children and nature is a recent extension of previous work by the author of varying aspects of human relationships to nature, most particularly perceptions, interactions, and behaviors relating to biological diversity (Kellert, 1996). This work has focused on the formation of basic meanings people attach and benefits they derive from the natural world, and the way these values are shaped by the influence of learning, culture, and experience, despite their presumed biological origins. The role of learning and maturation in childhood eventually emerged as a consideration in this examination.

In embarking on a review of scientific literature related to this subject, I soon discovered the paucity of systematic study of the role played by childhood contact with natural systems in character and personality formation. Two widely cited publications, for example, with the suggestive titles *The Ecology of Human Development* (Bronfenbrenner, 1979) and *Ecological Psychology* (Barker, 1968), almost exclusively employ the terms *ecology* and *environment* to consider family relationships, human social contexts, and the built rather than natural environment. These and other publications rarely considered the child's experience of the nonhuman world and its possible effects on physical and mental development. This omission became especially ironic when a doctoral student of mine asked a well-known developmental psychologist in my university to recommend relevant scientific literature on the subject. He was informed: "That's a very interesting question. I really can't recommend very much. I wonder why people haven't explored this subject. I'd be interested to learn what you discover. You know, you might talk with a professor at the Forestry School, Stephen Kellert, who I am certain could recommend extensive reading on the topic." This chapter may, thus, be something of "the blind leading the blind." One also wonders if the relative absence of published material on this subject may be indicative of a society so estranged from its natural origins it has failed to recognize our species' basic dependence on nature as a condition of growth and development.

A Conceptual Framework

A logical starting point in considering the potential impact of contact with nature in childhood development is to distinguish among kinds of experience children have with natural systems and processes. Young people's experience of nature, broadly speaking, can be classified in three ways—direct, indirect, and what may be called "vicarious" or "symbolic" experience (Kellert, 1996). *Direct experience* involves actual physical contact with natural settings and nonhuman species. The perspective adopted here, however, restricts these direct encounters to creatures and environments occurring largely outside and independent of the human built environment—that is, plants, animals, and habitats that function, for the most part, apart from continuous human input and control. The child's direct experience of nature is viewed as largely unplanned rather than formally organized into structured programs and activities, the latter included in a second category of indirect experience. Direct contact, thus, involves a young person's spontaneous play or activity in a backyard, in a nearby forest, meadow, creek, or neighborhood park, or even in an abandoned lot. In each case, the natural setting, though influenced by human manipulation and activity, includes creatures and habitats that function largely independent of human intervention and control.

A child's *indirect experience* of nature involves actual physical contact but in far more restricted, programmed, and managed contexts. Indirect experience of natural habitats and nonhuman creatures is typically the result of regulated and contrived human activity. Nature in these situations is usually the product of deliberate and extensive human mastery and manipulation. Examples might include children encountering plants, animals, and habitats in zoos, aquariums, botanical gardens, arboretums, natural history and science museums, and nature centers. A related but different expression of children's indirect experience of nature involves domesticated animals, plants, and habitats, particularly organisms and settings treated as an integral part of a child's home or family life. These domesticated forms include the "true companion" animals like cats and dogs, sometimes horses and birds, but also organisms that retain their "essential wildness" such as aquarium fish or potted plants. Indirect experience further includes contact with flower and vegetable gardens, cultivated crops and orchards, and domesticated farm animals, all habitats and creatures dependent on extensive human intervention and control.

Finally, *vicarious or symbolic experience* occurs in the absence of actual physical contact with the natural world. What the child encounters instead are representations or depicted scenes of nature that sometimes are realistic but that also, depending on circumstance, can be highly symbolic, metaphorical, or stylized characterizations. These vicarious images and symbolic depictions often occur in modern society through relatively innovative communication technologies like television, film, or computers, although more traditional print media such as books and magazines continue to be important conveyors of images of nature for children. Young people today clearly encounter an extraordinary array of vicarious images of nature. Yet the depiction of the natural world through symbolic means is ancient, perhaps as old as the human species itself, and is found extensively in cave and rock paintings, myths, totems, legends, and tales throughout human history (Jung, 1964; Levi-Strauss, 1966; Shepard, 1978, 1996).

This ancient lineage counters the inclination to treat vicarious treatments of nature as a particularly modern and possibly problematic aspect of contemporary society. What may be new today is the extraordinary proliferation of vicarious images and unprecedented technologies for representing nature through the mass media. Moreover, evidence suggests a concurrent decline in children's direct experience of healthy and abundant natural systems. In other words, what deeply worries some (Nabhan & Trimble, 1994; Pyle, 1993, see Chapter 11 in this volume by Orr and Chapter 12 by Pyle) is the contemporary erosion of direct and spontaneous contact with relatively undisturbed nature, especially among urban and suburban children, and a corresponding substitution of more artificial and symbolic encounters. Later in the chapter, the importance of children's direct encounters with relatively healthy and diverse natural systems in childhood maturation is considered, as well as the possible inadequacy of increasing indirect and vicarious experience as a developmental substitute. Considerable uncertainty will, nonetheless, be acknowledged regarding the functional and adaptive balance of varying degrees of direct, indirect, and vicarious contact with nature in child development.

A comprehensive theory of nature in childhood personality and character formation requires linking these three levels of experience with varying modes of learning in childhood development—cognitive, affective, and evaluative (values-related) maturation and development. *Cognitive or intellectual development* is broadly regarded here as emphasizing the formation of thinking and problem-solving skills; *affective maturation* as focusing on the emergence of emotional and feeling capacities; and *evaluative development* as stressing the creation of values,

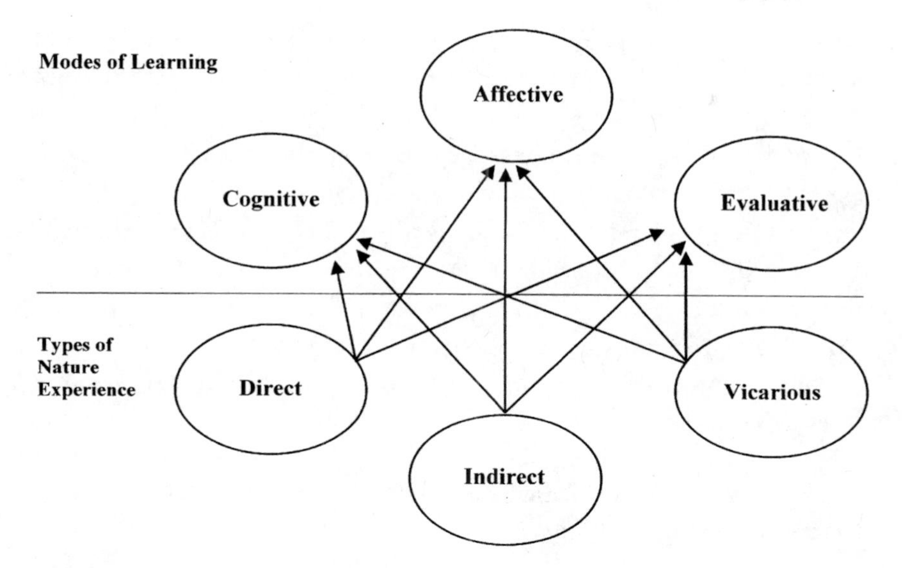

Figure 5.1
Types of Nature Experience and Modes of Learning in Childhood Development.

beliefs, and moral perspectives. The schematic relation of varying experiences of nature to cognitive, affective, and evaluative development is depicted in figure 5.1, and this framework guides much of the discussion in this chapter.

Cognitive Development

The taxonomy of cognition developed by Benjamin Bloom and colleagues (Bloom et al., 1956; Maker, 1982) can be usefully employed to explore the possible impact of varying forms of experience of nature in children's intellectual development. This taxonomy identifies six stages of cognitive maturation, moving broadly speaking from simple to complex levels of hierarchical and presumably sequential intellectual and problem-solving capacity:

· *Knowledge* Understanding facts and terms and applying this knowledge to the articulation and presentation of ideas, developing broad classificatory categories and systems, and recognizing of causal relationships;

• *Comprehension* Interpreting and paraphrasing information and ideas and extrapolating these understandings to other contexts and circumstances;
• *Application* Applying knowledge of general concepts, ideas, and principles to various situations and circumstances;
• *Analysis* Examining and breaking down knowledge into elements and categories and discerning underlying structural and organizational relationships;
• *Synthesis* Integrating and collating parts or elements into patterned, organized, and structural wholes and identifying and generating understandings of relationships and interdependencies; and
• *Evaluation* Rendering judgments about the functional significance and efficacy of varying elements and functions based on careful examination of evidence and impacts.

Limited evidence (Altman & Wohlwill, 1978; Kahn, 1999; Ratanapojnard, 2001) suggests that experiential contact with nature can exert a significant impact on cognitive development, especially during middle childhood and early adolescence. Insufficient space precludes, however, a thorough consideration of how this might occur for all six stages of cognitive development and in relation to each kind of nature experience, and, thus, only examples drawn from the knowledge and comprehension stages are briefly explored here by way of illustration.

A major focus of *knowledge* formation is developing basic understandings of facts and terms, creating categories and crude classification systems, and identifying causal relations. Limited empirical evidence (Kellert, 1996; Shepard, 1978) suggests that identifying, naming, classifying, and learning about the natural world can greatly facilitate the developing capacity for sorting and retaining information and ideas. At a relatively early age, children engage in the challenging and stimulating task of discerning, labeling, and classifying rudimentary features and behaviors of the natural environment. For example, in just about any environmental context, including the modern suburb and city, children encounter many and varied opportunities for naming and rudimentarily categorizing trees, birds, bushes, flowers, mammals, geological forms, and many other features of nature. Part of this appeal and function stems from the extraordinary variability of subject matter involved—in North America, for example, robins, cardinals, oaks, elms, rivers, streams, valleys, ants, mosquitoes, rocks, cliffs, boulders, and a plethora of vicariously encountered creatures such as giraffes, lions, tigers, bears, dinosaurs, and, of course, much more. The scale of accomplishment reflects the functional and adaptive significance associated with this wealth of opportunity for naming, sorting, and intellectual processing. The child confronts in nature an especially salient, diverse, and invigorating stream of objects and subjects useful in developing and practicing the capacities for labeling, differentiating, and classifying that are so basic to the knowledge stage of cognitive maturation.

The vicarious or symbolic experience of nature is an important aspect of this as well as other stages of cognitive development, although often insufficiently appreciated and recognized. A particularly interesting example is preschool books and stories (Kellert, 1996), where a content analysis of a random sample of these publications revealed an overwhelming preponderance of representations and images drawn from nature, particularly animals, used mainly for the purpose of nurturing the capacities for naming, classifying, and counting. Typically encountered were the equivalent of one bear, two giraffes, three lions, four hippos, and rarely were representations derived from a solely human-constructed world, such as one box, two pencils, three pins, four filing cabinets, five telephone poles. Many of these characterizations of nature were highly anthropomorphic and directed at emotional as much as intellectual maturation, but their association with naming and categorizing suggests their likely importance in facilitating the knowledge stage of cognitive development.

The anthropologist Elizabeth Lawrence (1993) employed the provocative term "cognitive biophilia" to underscore how symbols and images of nature are often used to facilitate human intellectual development. Even in a modern world of pervasive human domination and artificial construction, nature continues to provide young people with an unrivaled source of attraction, stimulation, and challenge relevant in both intellectual and emotional development. Edward O. Wilson (1993) suggested that the natural world is the most information-rich environment people will ever encounter. Moreover, this raw material is available to all children on an ongoing and spontaneous basis in relative proximity to where they reside and in more symbolic and indirect ways. What is uncertain today is how much the possible erosion of opportunities for direct contact with healthy and diverse natural environments has altered or compromised this basis for children's physical and mental development. The data on this issue are sparse and speculative and are reviewed later in the chapter. It would appear, however, that most children today, even those residing in highly populated areas, still confront a wealth of creative opportunities for experiencing nature in ways that continue to facilitate adaptive cognitive development. These experiences include direct encounters in backyards, parks, streetscapes, and more; indirect experiences through visiting zoos, natural history museums, nature centers, school programs, and so on; and vicarious contacts through proliferating electronic and print media, including magazines, books, film, television, and Web sites.

The role of contact with the natural world in the *comprehension* stage of cognitive development is briefly reviewed here before I examine analogous effects on affective and evaluative maturation. Comprehension broadly entails the translation, interpretation, and extrapolation of facts and ideas. In developing this capacity, the child systematically and relationally collates factual understanding with empirical experience. Encounters with the natural world, both real and imagined, provide a readily accessible context for this assimilation, analysis, and comprehension. The world that the child encounters includes such phenomena as snow falling at only certain temperatures; trees growing in particular climatic conditions but not in other conditions; ducks and geese being found in certain habitats but not in other habitats; butterflies appearing during the day but moths at night; large groups of diverse trees rather than small clumps of trees constituting a forest; herds of cattle and perhaps zebra occupying large expanses of grassland; crabs and clams occurring in wet and marshy areas rather than in dry and upland areas. The child confronts, in effect, nearly limitless contexts and opportunities in nature for developing and practicing the act of comprehension.

One could even suggest that few areas of life provide young people with as much opportunity as the natural world for critical thinking, creative inquiry, problem solving, and intellectual development (e.g., Berg & Medich, 1980; Chawla, 1988; Hart, 1979; Kahn, 1999, 1997; Kaplan & Kaplan, 1989; Kellert, 1997; Moore, 1986; Moore & Young, 1978; Nabhan & Trimble, 1994; Pyle, 1993; Searles, 1959; Sobel, 1993; Thomashaw, 1995). The challenging tasks of distinguishing one creature and one natural object from another, lumping life and nonlife into categories of relational division, observing the imperatives of feeding, surviving, reproducing, and dying: all and more offer a steady stream and rich diet for cognitive development. An intellectual adventure is facilitated by myriad observations and experiences of events and regularities in the natural world. Simple identification and classification of natural phenomena are followed by more complex conceptualizations, even principles yielding expectations and predictions of how nature behaves under varying circumstances and conditions. A process of intellectual competence spirals upward through a matrix of direct, indirect, and vicarious experiences of nature, strengthening the cognitive muscle we call mind and developing and reinforcing the child's capacities for empirical observation, analytical examination, and evidentiary demonstration.

Affective Development

Analogous effects can be identified in children's affective and values-related development in middle childhood and adolescence. A taxonomy of affective maturation has been devised by David Krathwohl, B. S. Bloom, and B. B. Masia (1964) and can be usefully employed in this regard. Five stages of emotional development have been identified by this formulation:

- *Receiving* Being aware and sensitive to facts and situations involving attentiveness and willingness to receive information;
- *Responding* Reacting and gaining satisfaction from receiving information and responding to situations;
- *Valuing* Attributing worth or importance to information and situations that reflect clear and consistent preferences and commitments;
- *Organizing* Internalizing and organizing preferences and assumptions of worth and importance into consistent patterns and sets of values and beliefs; and

• *Characterization by a value or value complex* Holding general patterns or sets of beliefs and values that constitute a coherent and consistent worldview or philosophy of life.

The Krathwohl et al. taxonomy includes both affective development and what is regarded here as evaluative or values-related maturation. Affective development is treated in this chapter as largely focusing on children's feelings and emotions, while values are regarded as a distinct synthesis of affective and cognitive perceptions and understandings (Kellert, 1996). Values are, in other words, viewed as an "emergent" phenomenon, a separate level of human experience involving the convergence of emotion and intellect that cannot be reduced to one or another of these states.

As indicated, Krathwohl's first stage of affective development focuses on a child's inclination to receive and respond to information and situations. The child's willingness to engage and interact with the world greatly depends on emotions such as like versus dislike, attraction as against aversion, wonder versus indifference, excitement in opposition to doubt, joy in place of sorrow, curiosity instead of boredom, courage versus fear, and more. In many if not most learning situations, affect precedes intellect as a basis for maturation and development. As Iozzi (1989a, 1989b, p. 5) suggests, "Significant evidence [exists] that the affective domain is the key entry point to learning and teaching."

What features and elements of a child's world typically foster and facilitate the inclination to receive information, to learn, and to develop? The support and security offered by parents, siblings, friends, teachers, and community are certainly the critical and irreplaceable core of this emotional foundation. But what about the subtle and complex matrix of interactions a child experiences with nature? The limited evidence available (Cornell, 1979; Derr, 2001; Kellert, 1985, 1996; Ratanapojnard, 2001; Sobel, 1993) suggests that contact with the natural world, especially during middle childhood, occupies a surprisingly important place in a child's emotional responsiveness and receptivity.

It is interesting to note that adult memories of childhood often emphasize the emotional significance of experiential relations to nearby nature that seemingly become a legacy carried into adulthood. For example, in her seminal study *The Ecology of Imagination in Childhood*, Edith Cobb (1977) describes many of these adult recollections. Based on these recalled encounters in nature in childhood, she concludes: "The child's ecological sense of . . . nature is . . . basically aesthetic and infused with joy in the power to know and to be. These equal, for the child, a sense of the power to make" (ibid., p. 23). Delight, elation, and affective engagement appear to form a crucible in which the child and later adult shape and mold an inclination for creativity and discovery. The emotive power of these encounters with nature derives from their dynamic, varied, often unique, surprising, and adventurous character. As Cobb surmises: "The child's sense of wonder, displayed as surprise and joy, is aroused as a response to the mystery of [the] stimulus [of nature] that promises 'more to come' or, better still, 'more to do'—the power of perceptual participation in the known and unknown" (ibid., p. 28). Invoking the poetic insights of Walt Whitman (Untermeyer, 1949, p. 346), Cobb suggests that the variety and diversity of the natural world nurture the child's capacities for creativity, beauty, and identity:

There was a child went forth every day,
And the first object he looked upon, that object he became,
And that object became part of him for the day or a certain part of the day,
Or for many years or stretching cycles of years.
The early lilacs became part of the child,
And grass and white and red morning glories, and white and red clover, and the song of the phoebe-bird,
And the Third-month lambs and the sow's pink-faint litter, and the mare's foal and the cow's calf.

The environmental scientist and writer Rachel Carson (1998) similarly observed that the child's emerging sense of joy and wonder is greatly enhanced by the emotional salience of nature. Carson found enthusiasm and passion for life, emotions essential to learning and personality formation, greatly benefiting from immersion and creative interaction with the diversity and mystery of nature. Emphasizing the importance of feelings toward the natural world as antecedent to intellectual growth, Carson (1998, p. 56) suggested:

For the child . . . , it is not half so important to *know* as to *feel*. If facts are the seeds that later produce knowledge and wisdom, then the emotions and the impressions of the senses are the fertile soil in which the seeds must grow.

> The years of early childhood are the time to prepare the soil. Once the emotions have been aroused—a sense of the beautiful, the excitement of the new and the unknown, a feeling of sympathy, pity, admiration or love—then we wish for knowledge about the object of our emotional response. Once found, it has lasting meaning. It is more important to pave the way for the child to want to know than to put him on a diet of facts he is not ready to assimilate.

The child's experience of nature encompasses a wide complex of emotions—wonder, satisfaction, joy, for sure, but also challenge, fear, and anxiety as well. The natural world provokes pleasure and enthusiasm but also a sense of uncertainty, danger, and at times terror. From the perspective of maturation and growth, all these and other emotions associated with the child's experience of nature serve as powerful motivators and stimuli for learning and development. As Rachel Sebba suggests (1991, p. 415), the natural world provides children with "an unfailing source of stimulation." The extraordinary conditions and characteristics in nature that provoke these diversity of emotional responses are arguably unique and nonduplicable. As the psychiatrist Harold Searles suggests (1959, p. 117), the child's relation to nature offers a directness and focus often lacking in human relationships: "The non-human environment is relatively simple and stable, rather than overwhelmingly complex and ever shifting . . . and generally available rather than walled off by parental injunctions."

Nature's emotional salience for the child also derives from its role in fantasy and imagination as much as from direct, literal, or tactile contacts. Nature is profoundly populated with creatures and habitats occurring in the realm of children's stories, myths, fables, tales, and dreams. These encounters provide a multitude of affective opportunities for engagement, discovery, creativity, joy, wonder, revelation, adventure, surprise, and more. Yet a natural world of imagination and vicarious experience can be emotionally dysfunctional if not balanced by contact with the actual and real of ordinary and everyday life. A worrisome feature of contemporary society is that many children increasingly experience nature through the imagined and exotic rather than through the actual and local. The functional and adaptive balance of varying kinds of contact with nature and the effects of possible declines in children's direct experience are considered later in the chapter. For now, it is suggested the spectacular and imagined in nature (the great African wildlife herds on television, the fantasy landscapes of Disney, the creatures of the museum and zoo) while stimulating do not substitute developmentally for more ordinary and everyday encounters with the natural world. The storyteller Dylan Thomas (1965, pp. 5, 8) describes the emotional adventure of directly experiencing nature in even mundane and familiar settings:

> Though it was a little park, it held within its borders of old tall trees, notched with our names and shabby from our climbing, as many secret places, caverns and forests, prairies and deserts, as a country somewhere at the end of the sea. . . . And though we could explore it one day, armed and desperate, from end to end, from the robbers' den to the pirates cabin, the highwayman's inn to the cattle ranch, or the hidden room in the undergrowth, where we held beetle races, and lit the wood fires and roasted potatoes and talked about Africa, and the makes of motor cars, yet still the next day, it remained as unexplored as the Poles. . . . And that park grew up with me; that small world widened as I learned its secrets and boundaries, as I discovered new refuges and ambushes in its woods and jungles; hidden homes and lairs for the multitudes of imagination.

Evaluative (Values-Related) Development

Finally, there is the consideration of the relation of values of nature to childhood development. Previous research (Kellert, 1996) has yielded a typology of nine basic values of the natural world. These values are thought to constitute "weak" biological tendencies or genetic inclinations to affiliate with natural process and diversity and are collectively labeled *biophilia* (Kellert & Wilson, 1993; Kellert, 1997; Wilson, 1984). As biological tendencies, these values reflect affinities for nature that presumably have proven adaptive in human evolution. As *weak* biological inclinations, the functional and adaptive occurrence of these values is viewed as greatly shaped by the mediating influence of learning, culture, and experience (Lumsden & Wilson, 1983). The content and intensity of the values vary greatly in individuals and groups, but this variability and its healthy and adaptive expression are seen as biologically limited and bounded. The insufficient and atrophied or the exaggerated and inordinate expression of any of the values is considered over the long-term dysfunctional and maladaptive.

Brief descriptions of the nine values of nature and their presumed adaptive significance to personality and character development in childhood are provided here. One-sentence definitions are also indicated in

Table 5.1
A Typology of Values of Nature

Value	Definition
Aesthetic	Physical attraction and appeal of nature
Dominionistic	Mastery and control of nature
Humanistic	Emotional bonding with nature
Moralistic	Ethical and spiritual relation to nature
Naturalistic	Exploration and discovery of nature
Negativistic	Fear and aversion of nature
Scientific	Knowledge and understanding of nature
Symbolic	Nature as a source of language and imagination
Utilitarian	Nature as a source of material and physical reward

table 5.1, although more thorough descriptions can be found elsewhere (Kellert, 1997).

The *aesthetic* value reflects the physical attraction and appeal of nature. Its development is viewed as instrumental in a child's emerging capacity for perceiving and recognizing order and organization, for developing ideas of harmony, balance, and symmetry, and for evoking and stimulating curiosity, imagination, and discovery.

The *dominionistic* value reflects the urge to master and control nature. Adaptive benefits associated with this value include safety and protection, independence and autonomy, the urge to explore and confront the unknown, and the willingness to take risks, be resourceful, and show courage.

The *humanistic* value emphasizes strong affection and emotional attachment for nature. Bonding with elements of the natural world is viewed as instrumental in developing intimacy, companionship, trust, capacities for social relationship, and affiliation and in enhancing self-confidence and self-esteem through giving, receiving, and sharing affection.

The *moralistic* value reflects an ethical and spiritual affinity for nature. Adaptive benefits associated with the formation of this value include a sense of underlying meaning, order, and purpose, the inclination to protect and treat nature with kindness and respect, and enhanced sociability from shared moral and spiritual conviction.

The *naturalistic* value expresses the desire for close contact and immersion in nature. Functional benefits stemming from this value include inclinations for exploration, discovery, curiosity, inquisitiveness, and imagination, enhanced self-confidence and self-esteem by demonstrating competence and adaptability in nature, and greater calm and coping capacities through heightened temporal awareness and spatial involvement.

A *negativistic* value reflects the avoidance, fear, and rejection of nature. Functional and adaptive benefits of this value include avoiding harm and injury, minimizing risk and uncertainty, and respect and awe of nature through recognizing its power to humble and destroy.

A *scientific* value emphasizes the empirical and systematic study and understanding of nature. Functional advantages of developing this value include intellectual competence, critical thinking, problem-solving abilities, enhanced capacities for empirical observation and analysis, and respect and appreciation for natural process and diversity.

The *symbolic* value indicates nature's role in shaping and assisting human communication and thought. Adaptive benefits of this value include classifying and labeling abilities instrumental in language and counting, resolution of difficult aspects of psychosocial development through story and fantasy, and enhanced communication and discourse through the use of imagery and symbol.

Finally, a *utilitarian* value reflects the material and commodity attraction of the natural world. Several advantages of this value include physical and material security, self-confidence and self-esteem through demonstrating craft and skill in nature, and recognition of human physical dependence on natural systems and processes.

Limited research (Eagles & Muffitt, 1990; Kellert & Westervelt, 1983; Kellert, 1985, 1996) suggests these nine values differentially emerge at varying ages or stages, somewhat analogous to Bloom's and Krathwohl's levels of cognitive and affective development (Lickona, 1991; Maker, 1982; Piaget, 1969). This developmental progression has four characteristics. First, like affective and cognitive development, formation of the values of nature typically moves from relatively concrete and direct perceptions and responses to more abstract levels of experiencing and thinking. Second, the values of nature generally shift from highly personal, egocentric, and self-centered concerns to more socially and other person interests. Third, the geographic focus of the values tends to move from local and parochial settings to more regional and global outlooks. Finally, predominantly emotional and affective values of nature emerge earlier than more abstract, logical, and rationally deduced perspectives.

For most children, the values of nature prominently develop at distinctive ages or stages (Kellert, 1996; Kellert & Westervelt, 1983). This typical developmental process does not suggest the absence or irrelevance of values at other times in a child's life but rather suggests periods when particular values prominently form and become manifest. For example, the formation of a moralistic value of nature is typically most pronounced and rapid during adolescence. This means not that substantially younger children lack the capacity or inclination to form moral perspectives of nature but that these views tend to be less developed and relevant at an earlier age, becoming most prominent and emergent at a later period or stage.

The first stage in the development of children's values of nature occurs between three and six years of age, emphasizing the formation of utilitarian, dominionistic, and negativistic perspectives of the natural world. This stage involves a primary emphasis on satisfying the child's material and physical needs, avoiding threat and danger, and achieving feelings of control, comfort, and security. Affection for nature, or the recognition of the independent needs of other creatures, is not absent, but these sentiments tend to be subordinated to more egocentric and personal needs and desires. Most children at this age show a pronounced indifference or anxiety toward direct contact with all but a small proportion of relatively familiar and domesticated creatures and natural settings.

A second developmental period in values of nature occurs during middle childhood from roughly six to 12 years of age. Middle childhood is a time when the humanistic, symbolic, aesthetic, and knowledge components of the scientific value develop most rapidly, while utilitarian, negativistic, and dominionistic perspectives diminish in importance.

Children at this age become more comfortable, familiar, and appreciative of other creatures and natural settings, although often in relative proximity to the home rather than in pristine or wild areas. Wildlife are more likely to be viewed as independent and autonomous, possessing feelings and interests apart from the child's personal needs and concerns. At this age, children better recognize the "different-ness and other-ness" of the natural world (Shepard, 1996), cultivating greater affection and curiosity for other creatures and environments. They are likely to venture into nonfamiliar natural settings, expanding their knowledge, sense of competence, and capacity to cope in these areas independent of adult supervision. They also emerge more cognizant of the autonomous rights of other life and begin to develop feelings of responsibility for care and considerate treatment of nature independent of being punished by adults for improper behavior. Most important, this is a time of greatly expanded interest, curiosity, and capacity for assimilating knowledge and understanding of the natural world. Rapid cognitive and intellectual growth occurs including many critical thinking and problem-solving skills achieved through interaction and coping in the nonhuman environment.

Intellectual development at this stage is especially facilitated by direct contact with nearby natural settings, where a world of exploration, imagination, and discovery becomes increasingly evident to the child. The importance of contact with nearby nature in personality formation during middle childhood has been emphasized by David Sobel (1993). Based on empirical studies of children's interactions with nature, Sobel concluded (1993, p. 159): "Middle childhood is a critical period in the development of the self and in the individual's relationship to the natural world. The *sense of wonder* of early childhood gets transmuted in middle childhood to a *sense of exploration*. Children leave the security of home behind and set out . . . to discover the new world." The psychiatrist Harold Searles (1959) also reported children at this age use nature to secure an identity apart from parents, the immediate home, and other people. Establishing familiarity in nearby natural environments and becoming constructive and creative in dealing with these settings provide a wealth of opportunities for generating feelings of autonomy, independence, and self-sufficiency. As Erik Erikson (1968) has suggested, middle childhood marks the time when children are especially interested in making things, in demonstrating industry and competence, and in establishing a self separate and apart from the continuous care and control of adults.

These objectives are often served by creating "places" outside although in proximity to the home—what children often refer to as forts, dens, secret hiding places, and the like. Sobel (1993, pp. 70, 74) explains the importance of the activity: "During this period of middle childhood, the self is fragile and under construction and needs to be protected. . . . The secretive nature of the hiding place is significant. . . . These places are valuable in that they provide the satisfaction of being able to transform the environment successfully and comfort in being able to make a place for oneself." These constructed and intimate places outside but near the home, secreted in the foliage of trees and bushes of ordinary nature, offer the child the chance to create and construct quasi-habitations. In the process, they achieve both autonomy and a surging confidence in demonstrating the ability to produce something from raw nature signifying both safety and security.

These places are also at the margins of the child's known world, providing considerable potential for exploration, discovery, and adventure. Moreover, these experiences can become deeply etched in the child's mind at this age, as Wallace Stegner (1962, p. 21) describes: "There is a time somewhere between five and twelve . . . when an impression lasting only a few seconds may be imprinted . . . for life. . . . Expose a child to a particular environment at his susceptible time and he will perceive in the shapes of the environment until he [or she] dies." These moments of intimate relation to the natural world often seem timeless, as Loren Eiseley (1946, p. 208) suggests: "Once in a lifetime . . . , one so merges with sunlight and air and running water that whole eons . . . pass in a single afternoon."

These direct encounters in nature do not suggest that indirect and vicarious experiences are developmentally unimportant or irrelevant at this age. As previously suggested, images and representations of the natural world that are instrumental in naming, labeling, classifying, and language acquisition are important at an earlier stage in child development. During middle childhood, continuing but different vicarious and symbolic contacts with nature assume significance. At this later age, children are especially enchanted by tales, legends, stories, and myths involving scenes and characters drawn from the natural world (Bettelheim, 1977; Engel, 1995; Kellert, 1997). These often fantastic portrayals of nature frequently involve plots focusing on fundamental issues of identity and selfhood—good against evil, freedom fighting tyranny, order versus chaos, innocence in the face of sexuality, and more. In these narratives, images of the natural world are often distorted. Yet these frequent anthropomorphic depictions of essentially animals in human disguise often help to render more tolerable the challenging developmental dilemmas of conflict, need, control, and desire. Paul Shepard (1996, p. 76) elucidates the possible maturational significance of these stories, invoking Bruno Bettelheim's (1977) seminal analysis:

> The fairy tale [myth, fable, legend, dream] dramatizes the intrinsic childhood worries which the youthful listener unconsciously interprets as his own story and his own inner self. . . . Bettelheim believes the problems to be universal, having to do with protection from malicious relatives, the uncertain intentions of strangers, [the child's] verbal or physical limitations such as the skills of speech and strength, the bodily changes and functions associated with growth, frightening dreams . . . , Oedipal feelings, sibling rivalry, jealousy and envy. . . . Every story is a magic prophecy of personal transcendence. . . . Their message is that special skills, often the powers represented by different animal species, will come to the rescue, solve the problems, save the day, and guarantee a happy lifetime.

These vicarious experiences of the natural world help the young person to navigate the perilous issues of maturation in a vivid, often mysterious, and unusually beguiling way. When coupled with direct contact and immersion in nearby nature, these symbolic encounters provide extraordinary opportunities for psychosocial growth and development.

A third and final stage in the development of values of nature occurs between 13 and 17 years of age. This period witnesses a rapid and pronounced maturation of more abstract, conceptual, and ethical reasonings about the natural world—in the terms of the values typology, a significant expansion in moralistic, naturalistic, and ecological components of the scientific values of nature. Adolescence is a time when children become cognizant and appreciative of larger spatial and temporal scales (such as ecosystems, landscapes, and evolutionary processes) that are difficult to visualize but indicative of vital human dependencies on natural systems. Adolescent children also reveal a more focused and conceptually complex understanding of ethical responsibilities for nature, including morally acceptable and legitimate treatment and behavior. They emerge more cognizant of the presumed experiences of nonhuman beings and more keenly aware of the capacity of other creatures to suffer. Ideas about nature thus become more abstract and systematic, including understandings of the complex relationships linking humans with the natural world, such as ecological and energy flows, as well as concepts of morality and environmental stewardship.

Most children at this age also engage in daring, expansive, and challenging activities testing the physical limits of the natural world. In doing so, they often nurture self-confidence, self-esteem, and an increased sense of identity. Opportunities for this kind of experience of nature in modern society often occur through participating in outdoor programs involving considerable challenge in relatively undisturbed and unfamiliar settings.

The potential impact on adolescent maturation of this kind of activity is suggested by a recent major study (Kellert & Derr, 1998) of programs offered by three well-known organizations: Outward Bound (OB), the National Outdoor Leadership School (NOLS), and the Student Conservation Association (SCA). Over 700 persons participated in two related studies—a retrospective investigation of persons who had participated in one of the programs during the past approximately 25 years and a longitudinal study of participants immediately before, just after, and six months following a program. Data-collection procedures included surveys, in-depth interviews, and observations. A selection of relevant findings is reviewed here, and their similarities are emphasized, although program differences also occurred that were largely indicative of varying organizational philosophies and activities. In both the retrospective and longitudinal studies, a large majority of participants reported the experience as being one of the most important in their lives and as having exerted major impacts on their personality and character development. Moreover, these views rather than diminishing slightly increased over time in both the retrospective study, many years following program participation, and in the longitudinal investigation, six months after participating. More specifically, the great majority of respondents indicated the experience had greatly enhanced their self-confidence, self-concept, and capacity to cope with adversity and challenge. Two-thirds to three-quarters also reported major and sustained improvements in self-esteem, independence, autonomy, initiative, decision-making and problem-solving abilities, and interpersonal skills and relationships. Most respondents further noted this primarily wilderness experience had positive effects on their capacity to function in more modern and urban settings. Finally, a large majority indicated greater respect and appreciation for nature, increased participation in various outdoor activities, and support for nature conservation.

More qualitative but powerful articulations of the perceived importance of this nature experience on character and personality development are provided by the following participant comments from Kellert and Derr (1998, p. 235):

> Participating [occurred at] a pivotal point in my life. It gave me the opportunity to take a risk. It strengthened my sense of self. It gave me a feeling of purposefulness, self-respect, and strength that I had never had before. When you have confidence in yourself, it affects every aspect of your life.

> It was the most amazing, awe-inspiring, thought provoking, and challenging experience of my life. . . . It helped me to believe that if there is anything I really want to do in life, I have the ability to do it. All I have to do is look deep inside myself, and I can find it. . . . [It] helped me realize who I was and how I fit into the world around me. This realization affects every decision I make in my life.

> The experience, while isolated and out of the realm of everyday life, is applicable to everything I do. Because everything in the wilderness was such raw emotion and the outer events so simple, the personal challenges faced and overcome were within myself. Much of what I faced . . . had to do with my own fears and weaknesses. Overcoming them changed me as a person. When I face a more "complex" problem in the outside world, I need only to go back to see what solution I came to when it was just me against myself surrounded by simplicity, and the answer becomes clearer. [It] allowed me to experience a connection with nature and simplicity and balance within that will be with me for the rest of my life.

> It gave me an unbelievable confidence in myself. I found a beauty, strength, and an inner peace that I never knew was present. . . . I learned the most I ever learned about life, myself, and skills that I still use everyday. . . . It made me more confident, focused, and self-reliant. I have become more compassionate toward not only nature but toward other people. . . . I learned about respect, setting goals,

working to my maximum and past it. These are skills I consider to be important in everything I do, and I feel they will help me continue to be successful throughout life.

Despite these testimonials and the statistical results, limited impacts and deficiencies were also encountered, such as minimal effects on factual knowledge of natural systems. The results, however, largely indicated that challenge and immersion in relatively undisturbed and unfamiliar natural settings can exert major positive character and personality development impacts during adolescence. This finding has also been corroborated by other research (e.g., Driver et al., 1987; Ewart, 1989; Kaplan & Talbot, 1983).

The Importance of Direct Experience

It appears that during adolescence, challenge in relatively unusual and undisturbed natural settings can exert major development impacts. In earlier discussions, however, it was suggested that for children in middle childhood access and opportunity to nearby natural areas play more significant maturational roles. Like a young plant whose vigorous growth depends on establishing firm roots in a particular locale, children during middle childhood seem to rely on settled and familiar terrain for various maturational purposes. The potential importance of building roots in particular places for young people has also been emphasized by the psychiatrist Robert Coles (1971, p. 116), who suggested: "It is utterly part of our nature to want roots, to struggle for roots, for a sense of belonging, for some place that is recognized as *mine*, as *yours*, as *ours*." Simone Weil (1955, p. 43) echoed this view when she remarked, "To be rooted is perhaps the most important and least recognized need of the human soul." Studies by Derr (2001), Moore (1986), Nabhan and Trimble (1994), Sobel (1993), and others have also found children during middle childhood frequently depend more on nearby and familiar places than experiences in relatively unusual and spectacular environments. The importance of contact with "ordinary" nature is powerfully suggested by Pyle (1993, pp. xv, xix): "It is through close and intimate contact with a particular patch of ground that we learn to respond to the earth. . . . We need to recognize the humble places where this alchemy occurs. . . .

Everybody has a ditch, or ought to. For only the ditches—and the fields, the woods, the ravines—can teach us to care enough."

What seems evident—whether focusing on relatively ordinary and familiar natural settings during middle childhood or more challenging and undisturbed environments in adolescence—is that direct experience of nature plays a significant, vital, and perhaps irreplaceable role in affective, cognitive, and evaluative development. More study, of course, is needed before this conclusion can be confidently accepted. Still, the conceptual and empirical material presented tentatively support this claim. As the psychiatrist Harold Searles (1959, p. 27) suggested more than three decades ago: "The non-human environment, far from being of little or no account to human personality development, constitutes one of the most basically important ingredients of human psychological existence." This hypothesis substantiates the intuition of most adults regarding the importance of contact with nature during childhood, as reported by Sebba (1991, p. 400): "Despite the heterogeneous nature of the [study] participants in terms of sex, age, character, and the environments in which they grew up, 96.5% of them indicated the outdoors was the most significant environment in their childhood."

What is it, then, about nature that so attracts, stimulates, and retains the child's attention to the degree that it appears to exert a significant effect on childhood maturation and development? This topic raises issues that extend beyond the scope of this chapter and are somewhat considered by other chapters in this book. Still, several important characteristics can be noted to explain the particular allure and significance of the natural environment during middle childhood and early adolescence, drawing mainly on the work of Sebba (1991).

Sebba (1991, p. 416) suggests that an especially compelling feature of nature is that "the stimuli of the natural environment . . . assault the senses at an uncontrolled strength." For most if not all children, the extraordinary sensory diversity and variability of the natural world is unavoidable. Young people are surrounded and immersed, even in most urban settings, by a multiplicity of sights, sounds, smells, and tactile stimuli originating from the natural environment, whether explicitly or consciously recognized or not. Sebba (1991, p. 417) also suggests, "The natural environment is characterized by a continual change of stimuli (over time or across area)." In other words, the sensory effects of nature on the child tend to shift repeatedly in relation to continuous spatial and temporal modulation increasing the likelihood of the child's awareness, recognition, and response. Sebba (1991, p. 417) further notes, "Compared to the [human] built environment, the external environment is characterized by instability, which requires alertness and attention." Not only is the child confronted by continuous change in nature, but these alterations are often unpredictable and challenging, necessitating a wide range of adaptive and problem-solving responses.

Finally, Sebba (1991, p. 417) emphasizes the child's realization that "the natural environment is one from which life springs and one which exerts forces that cause inanimate objects to move." This formulation underscores the arguably most powerful, intense, and meaningful attraction of the natural world for the maturing child: it is replete with life and lifelike features and processes. Nature is intrinsically and qualitatively different from anything the child confronts in the human built world, no matter how well simulated, technologically sophisticated, or "virtual" these manufactured representations may be. Nature for the child fundamentally signifies life, a riot of distinctive and unique organisms that move, grow, reproduce, and often seemingly feel and think. Even nonliving elements—water, soil, rocks, geological forms, the atmosphere—typically strike the child as quasi-living entities, not exactly alive but frequently recognized as supporting and bringing forth life. The child is not an ecologist, but he or she can discern how life relies on clean and abundant water, plants grow in soil, animals eat plants and sometimes other animals, and the air moves like some great ambient beast and is the unavoidable basis of life. Eloquently reflecting on the developmental significance of these unique dimensions of nature, Rachel Carson (1998, pp. 54, 100) remarked:

> A child's world is fresh and new and beautiful, full of wonder and excitement. . . . What is the value of preserving and strengthening this sense of awe and wonder, this recognition of something beyond the boundaries of human existence? Is the exploration of the natural world just a pleasant way to pass the golden hours of childhood or is there something deeper? I am sure there is something much deeper, something lasting and significant. . . . Those who contemplate the beauty of the earth find reserves of strength that will endure as long as life lasts. There is symbolic as well as actual beauty in the migration of the birds,

the ebb and flow of the tides, the folded bud ready for the spring. There is something infinitely healing in the repeated refrains of nature.

The child's experience of nature is, thus, portrayed as an essential, critical, and irreplaceable dimension of healthy maturation and development. Assuming this to be so, the question raised earlier confronts us once again: Does modern society provide sufficient quantity and quality of opportunities for youthful experience of the natural world? Some researchers and writers (see Chapter 11 in this volume by Orr and Chapter 12 by Pyle) suggest that our age has witnessed greatly diminished and compromised possibilities for satisfying interaction between young people and nature. Pyle (1993) powerfully addresses this possibility in his concept of "the extinction of experience," which can be usefully examined in exploring the claim that modern society has eroded and impoverished childhood opportunities for adequate contact with the natural world. Pyle (1993, pp. 145, 147, italics added) describes the notion of "extinction of experience":

> Simply stated, the loss of neighborhood species endangers our experience of nature. . . . Direct, personal contact with living things affects us in vital ways that vicarious experience can never replace. I believe that one of the greatest causes of the ecological crisis is the state of personal alienation from nature in which many people live. We lack a widespread sense of intimacy with the living world. . . . The *extinction of experience* . . . implies a cycle of disaffection that can have disastrous consequences. As cities and metastasizing suburbs forsake their natural diversity, and their citizens grow more removed from personal contact with nature, awareness and appreciation retreat. . . . So it goes . . . the extinction of experience sucking the life from the land, the intimacy from our connections.

What empirical evidence exists to support the claim that contemporary children confront substantially fewer opportunities for direct and spontaneous contact with relatively familiar natural settings? Additionally, one wonders if possible increases in indirect and vicarious experiences of nature in modern society have perhaps substituted or compensated for declines in more direct encounters?

Data (Barney et al., 1980; Groombridge, 1992; Heywood, 1995; Myers, 1994; Savage, 1995; Wilcove et al., 1998; Wilson, 1992) relevant to the first question appear to corroborate an hypothesized "extinction" or at least a greatly diminished experience of nature among children today. Various dimensions of contemporary environmental degradation and decline—extensive habitat destruction, species loss, environmental contamination, natural resource depletion, urban sprawl, human population growth—all point toward substantially fewer opportunities for most children, especially in densely populated areas, to have contact with high-quality natural environments.

Massive contemporary declines in biological diversity, for example, constitute a virtual "hemorrhaging" of life on earth with an estimated 15,000 to 30,000 species extinctions occurring annually and pronounced declines in many species short of biological elimination (Wilson, 1992; Kellert, 1997). On the other hand, while this biological decline is lamentable and ominous, does it equate with an "extinction of experience"? Most of the species that now face the greatest threat of extinction are obscure and unknown invertebrates of the remote tropics, and most of them occur in the forest canopy. Yet pronounced reductions in biological diversity also reflect widespread population reductions and associated loss and simplification of habitats and ecosystems familiar to children. Moreover, extensive urban sprawl, building of massive road and vehicular networks, and widespread conversion of natural to artificial environments have resulted in pervasive loss of common species and habitats.

Most important, the great majority of contemporary children continually confront these symptoms of environmental decline. They have also borne witness to widespread reductions and extirpations of many of the charismatic (the "phenomenologically" salient) fauna and flora—the grizzly bears, whales, tigers, pandas, lions, elephants, redwoods, and more (Kellert, 1997). One wonders about the consequence of having so many traditional symbols of awe, wonder, and beauty in nature become ubiquitous signs of rarity, loss, and decline. But what especially concerns Pyle and others is the loss of accessible, spontaneous, and challenging encounters with familiar, nearby, and "everyday" nature for youth today. They lament the elimination, fragmentation, isolation, and contamination of pockets of naturalness once characteristic of most neighborhoods and communities, even in urban areas. Moreover, they bemoan how the remaining habitats so often become victims of invasion and replacement of native by nonnative organisms, further signifying not just ecological decline but also the loss of historically familiar nature.

Pronounced changes in the social and cultural contexts of most children's experiences of the natural world have also exacerbated this impoverished condition. Major shifts in family traditions, recreational activity, social support networks, and community relations have eroded many children's traditional opportunities for contact with nature. Studies have demonstrated that adults often perform the critical role of encouraging a child's interest and experience of nature, recruitment and commitment to various outdoor activities often depending on shared cultural traditions passed from one generation to another (Applegate, 1991; Carson, 1998; Duda & Young, 1994; Nabhan & St. Antoine, 1993). Increased mobility and transience, the shift from a norm of extended to nuclear families, and the erosion of stable communities and a sense of place have all resulted in substantial reductions in opportunities for children and adults to share and cultivate these experiences of nature. Growing concerns about the safety of children to function independent of adult supervision and increasing dependence on vehicular transportation constitute further obstacles to spontaneous and familiar interactions with nearby nature among many contemporary children.

What about the second question: Could possible increases in indirect and vicarious experiences of nature in modern society have substituted or compensated for declines in more direct encounters with the natural world? More children today than ever appear to participate in formally organized indirect activities involving animals and nature offered by schools, nature centers, outdoor programs, and visits to zoos, natural history and science museums, and botanical gardens. Moreover, contemporary children have unprecedented and revolutionary access to nature through the technological inventions of television, film, video, computers, the Internet, and other electronic media. As a consequence of these changes, children experience far greater exposure to natural settings and creatures than could have been imagined previously, and the likelihood is that similarly revolutionary technical changes will occur in the future.

Many (e.g., Kellert, 1997; Mander, 1991; Nabhan & Trimble, 1994; Pyle, 1993) have nonetheless questioned the degree and importance of learning and developmental effects associated with these vicarious experiences of nature, especially when they occur in a context of greatly diminished and declining direct contacts with familiar, healthy, and accessible habitats. These critics argue that encounters with nature on television, on film, or through the computer can never provide the challenge, immersion, intimacy, discovery, creativity, adventure, surprise, and more afforded by direct and spontaneous experiences in familiar natural settings. Pyle (1993, p. 146) remarked in this regard, "Everyone has . . . a chance of realizing a pleasurable and collegial wholeness with nature. But to get there, intimate association is necessary. A face-to-face encounter with a banana slug means much more than a Komodo dragon seen on television. . . . Direct, personal contact with other living things affects us in vital ways that vicarious experience can never replace."

What about the value and impact of increased indirect contact with nature through more frequent visits to high-quality zoological parks, aquariums, natural history museums, outdoor programs, and nature centers? Insufficient data preclude confident conclusions regarding whether these more frequent indirect encounters with nature compensate for declines in direct experience, although some suggestive results can be provided from studies of zoos and outdoor programs.

Zoos, it should be noted, are far from a contemporary invention. The first known collection of wild animals in captivity, in fact, occurred more than four thousand years ago, and the modern zoo developed in Europe during the sixteenth and seventeenth centuries. Today, more than 400 professionally managed zoological parks exist worldwide. Annual visits to zoos and aquariums in the United States total more than 100 million, with a remarkable 98 percent of Americans having visited a zoo at some point in their lives (Dunlap & Kellert, 1994).

Various studies (Birney, 1986; Kellert & Dunlap, 1989; Kellert, 1996; Kellert & Vollbracht, 2000) suggest that increasingly sophisticated naturalistic exhibits and ambitious educational programs at zoos and aquariums can exert major positive learning effects on children. Still, these impacts are typically transitory and unlikely to produce significant effects on children's character and personality development. Not only are zoo and museum visits typically infrequent, self-paced, unstructured, passive, and entertainment-oriented, but encounters with nature and animals in these informal learning settings often focus on unusual and rare rather than local or familiar creatures and habitats. Finally, no matter how sophisticated the exhibit, it still constitutes the attempt to simulate, fabricate, and re-create reality: it is fundamentally a "show" whose artificiality is recognized by most visitors, including children. Zoo and museum experiences lack the intimacy, challenge, creativity, and active participation afforded by more direct encounters with the natural world. Expanding opportunities for children to experience nature in zoos and other museum settings does not, thus, appear to provide sufficient or adequate substitution for declines in more direct and spontaneous encounters with the natural world.

On the other hand, study results reported earlier in the chapter indicate the remarkable and enduring learning and character development impacts of participation in outdoor programs and associated travel to relatively undisturbed natural areas. Several factors limit the potential developmental importance of these activities. First, only a small fraction of today's youth has the opportunity to participate in these outdoor programs. Second, their occurrence in unusual natural settings casts doubt on how much these experiences substitute or compensate for significant declines in more direct encounters with nature in ordinary and familiar environments. Third, participation in these outdoor programs and their likely impact seem limited to mainly older youth. Finally, the outdoor programs often involve activities very unlike the "normal" realities faced by most young people. As one participant remarked:

> [Participation] shifted my perspective a little bit. . . . I went and gathered some strength. But the experience now seems so distant. Everything we learned is relevant here, but it is so abstract. We learned how to organize and be careful with what we do with our bodies. But with the everyday hustle [and] bustle of daily life it is hard to incorporate this into my life.

It may be tentatively concluded that increases in children's indirect contact with nature in modern society do not exert major or long-term developmental impacts on most young people. A basic deficiency of most indirect activities is that their sporadic, atypical, highly structured, and planned features often limit the spontaneity and adaptive behavior provided by less restricted and managed encounters in the natural world. Outdoor programs, visits to zoos and museum-like settings, and other indirect experiences likely exert their most positive effects when they complement direct encounters in familiar natural environments. Echoing this conclusion about the value of youth visits to protected areas, Pyle (1993, p. 148) suggested, "Nature reserves . . . are not enough to ensure connections. Such places, important as they are, invite a measured, restricted kind of contact. . . . Children . . . need free places for puttering, netting, catching, and watching. . . . Spots near home where [they] can wander off a trail, lift a stone, poke about, and merely wander: places where no interpretive signs intrude their message to rob [their] spontaneous response."

The vital and seemingly irreplaceable importance of children's direct experience of nature in even modest and sometimes compromised natural settings is further suggested by the results of studies in diverse cultural and economic locations (Derr, 2001; Hart, 1979, 1997; Moore, 1986; Ratanapojnard, 2001; Sobel, 1993). These findings also indicate that direct encounters with nature provide children with unique and critical developmental opportunities for discovery, creativity, and personal autonomy. Moore (1986) in a study of English children encountered youth engaged in experientially rich activities in even small and remnant natural areas in urban and sometimes environmentally degraded settings. David Sobel (1993) similarly observed children create outposts of wonder, learning, and great personal satisfaction in relatively modest outdoor areas. Pyle (1993, pp. 148–149) articulates well the potential importance of these direct encounters: "Nothing serves better than the hand-me-down habitats that lie somewhere between formal protection and development. . . . Developers, realtors, and the common parlance refer to such weedy enclaves as 'vacant lots' and 'waste ground.' . . . I grew up in a landscape lavishly scattered with unofficial countryside. . . . They were rich with possibility."

Conclusion

This chapter concludes with some ambivalence. Both theory and data have provided tentative support for the hypothesis that children's emotional, intellectual, and values-related development, especially during middle childhood and early adolescence, is greatly enhanced by varied, recurrent, and ongoing contact with relatively familiar natural settings and processes. Moreover, these encounters benefit from a mix of direct, indirect, and vicarious or symbolic experiences of nature. The more somber conclusion is that various trends in modern society—unsustainable consumption, urban sprawl, biodiversity loss, chemical contamination—have resulted in pronounced and significant declines in the quality and quantity of children's direct experience of the natural world. Moreover, possible increases in children's indirect and vicarious contact with nature do not appear to offer an adequate substitute for diminished direct encounters in ordinary and accessible natural environments.

What can be done to correct or improve this presumed deficiency of many if not most children's socialization in modern society? This is a difficult and complicated subject extending far beyond the scope of this chapter. The good news is that a substantial fraction of youth today continue to encounter rich and rewarding opportunities for experiencing nature. On the other hand, the majority of contemporary children will not likely have contact with abundant and quality natural settings until fundamental changes occur in the activities and perspectives of most planners, educators, developers, leaders, and families. We require a radical shift in the ways we design and construct our homes, schools, recreational facilities, open spaces, and communities that deliberately seeks to incorporate all values of nature as an essential core of children's lives.

Acknowledgments

The author has been greatly influenced by many in addressing the complicated subject of the experience of nature in childhood development. I want to particularly thank three of my doctoral students, Victoria Derr, Sorrayut Ratanapojard, and Kevin Eddings, whose ideas and research have greatly assisted me in writing this chapter.

References

Altman, I., & Wohlwill, J. (Eds.). (1978). *Children and the environment*. New York: Plenum Press.

Applegate, J. (1991). Patterns of early desertion among New Jersey hunters. *Wildlife Society Bulletin, 17*, 476–481.

Barney, G. (Ed.). (1980). *Global 2000 report*. Washington, DC: Council on Environmental Quality.

Barker, R. (1968). *Ecological psychology: Concepts and methods for studying environment in human behavior*. Stanford, CA: Stanford University Press.

Berg, M., & Medrich, E. A. (1980). Children in four neighborhoods: The physical environment and its effect on play and play patterns. *Environment and Behavior, 12*, 320–348.

Bettelheim, B. (1977). *The uses of enchantment: The meaning and importance of fairy tales*. New York: Vintage Books.

Birney, B. (1986). *A comparative study of children's perceptions and knowledge of wildlife as they relate to field trip experiences at the Los Angeles County Museum of Natural History and the Los Angeles Zoo*. Ann Arbor: University Microfilms.

Bloom, B. S., Engelhart, M. B., Furst, E. J., Hill, W. H., & Krathwohl, D. R. (1956). *Taxonomy of educational objectives, Handbook I: The classification of educational goals—Cognitive domain*. New York: Longman.

Bronfrenbrenner, U. (1979). *The ecology of human development: Experiments by nature and design*. Cambridge, MA: Harvard University Press.

Carson, R. (1998). *The sense of wonder*. New York: HarperCollins.

Chawla, L. (1988). Children's concern for the natural environment. *Children's Environment Quarterly, 5*, 13–20.

Cobb, E. (1977). *The ecology of imagination in childhood*. New York: Columbia University Press.

Coles, R. (1971). *Migrants, sharecroppers, mountaineers*. Boston: Little, Brown.

Cornell, J. (1979). *Sharing nature with children*. Nevada City, CA: Dawn.

Derr, V. (2001). Growing up in the Hispano homeland: The interplay of nature, family, culture, and community in shaping children's experiences and sense of place. Doctoral dissertation, School of Forestry and Environmental Studies, Yale University.

Driver, B., et al. (1987). Wilderness benefits: A state-of-the-knowledge review. In R. C. Lucas (Ed.), *Proceedings of the National Wilderness Research Conference*, General Technical Report INT-220. Ft. Collins, CO: U.S. Forest Service.

Duda, M. (1993). *Factors related to hunting and fishing participation in the United States*. Harrisonburg, VA: Responsive Management.

Duda, M., & Young, K. 1994. *Americans and wildlife diversity*. Harrisonburg, VA: Responsive Management.

Dunlap, J., & Kellert, S. R. (1994). Zoos and zoological parks. In W. Reich (Ed.), *Encyclopedia of bioethics*. New York: Macmillan.

Eagles, P. J., & Muffitt, S. (1990). An analysis of children's attitudes toward animals. *Journal of Environmental Education, 21*, 41–44.

Eiseley, L. (1946). *The immense journey.* New York: Random House.

Engel, S. (1995). *The stories children tell: Making sense of the narratives of childhood.* New York: Freeman.

Erikson, E. (1968). *Identity: Youth and crisis.* New York: Norton.

Ewart, A. (1989). *Outdoor adventure pursuits: Foundations, models and theories.* Scottsdale: Publishing Horizons.

Groombridge, B. (Ed.). (1992). *Global biodiversity.* London: Chapman & Hall.

Hart, R. A. (1979). *Children's experience of place.* New York: Knopf.

Hart, R. A. (1997). *Children's participation: The theory and practice of involving young citizens in community development and environmental care.* London: Earthscan.

Heywood, V. (Ed.). (1995). *Global biodiversity assessment.* Cambridge: United Nations Environment Program/Cambridge University Press.

Iozzi, L. A. (1989a). What research says to the educator, Part 1, Environmental education and the affective domain. *Journal of Environmental Education, 20,* 3–9.

Iozzi, L. A. (1989b). What research says to the educator, Part 2, Environmental education and the affective domain. *Journal of Environmental Education, 20,* 6–13.

Jung, C. (1964). *Man and his symbols.* Garden City, NJ: Doubleday.

Kahn, P. H., Jr. (1997). Developmental psychology and the biophilia hypothesis: Children's affiliation with nature. *Developmental Review, 17,* 1–61.

Kahn, P. H., Jr. (1999). *The human relationship with nature: Development and culture.* Cambridge, MA: MIT Press.

Kaplan, S., & Kaplan, R. (1989). *The experience of nature.* New York: Cambridge University Press.

Kaplan, S., & Talbot, J. (1983). Psychological benefits of a wilderness experience. In I. Altman & J. Wohlwill (Eds.), *Behavior and the natural environment.* New York: Plenum Press.

Kellert, S. R. (1985). Attitudes toward animals: Age-related development among children. *Journal of Environmental Education, 16,* 29–39.

Kellert, S. R. (1996). *The value of life: Biological diversity and human society.* Washington, DC: Island Press.

Kellert, S. R. (1997). *Kinship to mastery: Biophilia in human evolution and development.* Washington, DC: Island Press.

Kellert, S. R., & Derr, V. (1998). *National study of outdoor wilderness experience.* New Haven: Yale University School of Forestry and Environmental Studies.

Kellert, S. R., & Dunlap, J. (1989). *Informal learning at the zoo.* New Haven: Yale University School of Forestry and Environmental Studies.

Kellert, S. R., & Vollbracht, C. (2000). *A review of literature and research relating to the learning impacts on children of zoological parks and aquariums.* New Haven: Yale University School of Forestry and Environmental Studies.

Kellert, S. R., & Westervelt, M. O. 1983. *Children's attitudes, knowledge and behaviors toward animals (Phase 5).* Washington, DC: U.S. Fish & Wildlife Service.

Kellert, S. R., & Wilson, E. O. (Eds.). (1993). *The biophilia hypothesis.* Washington, DC: Island Press.

Krathwohl, D. R., Bloom, B. S., & Masia, B. B. (1964). *Taxonomy of educational objectives: The classification of educational goals*, Handbook 2, *Affective domain.* New York: Longman.

Lawrence, E. (1993). The sacred bee, the filthy pig, and the bat out of hell: Animal symbolism as cognitive biophilia. In S. Kellert & E. O. Wilson (Eds.), *The biophilia hypothesis.* Washington, DC: Island Press.

Levi-Strauss, C. (1966). *The savage mind.* Chicago: University of Chicago Press.

Lickona, T. (1991). *Educating for character.* New York: Bantam Books.

Lumsden, D., & Wilson, E. O. (1983). The relation between biological and cultural evolution. *Journal of Biological Structure, 8*, 343–359.

Maker, C. J. (1982). *Teaching models of the gifted.* Austin: Pro-ed.

Mander, J. (1991). *In the absence of the sacred.* San Francisco: Sierra Club Books.

Moore, R. C. (1986). *Childhood's domain: Play and space in child development.* London: Croom Helm.

Moore, R., & Young, D. (1978). Childhood outdoors: Toward a social ecology of the landscape. In I. Altman & J. F. Wohlwill (Eds.), *Children and the Environment* (pp. 83–130). New York: Plenum Press.

Myers, N. (1994). Global biodiversity II: Losses. In G. Meffe & C. Carrolls (Eds.), *Principles of conservation biology.* Sunderland, OH: Sinauer.

Nabhan, G. P., & St. Antoine, S. (1993). The loss of floral and faunal story: The extinction of experience. In S. R. Kellert & E. O. Wilson (Eds.), *The biophilia hypothesis* (pp. 229–250). Washington, DC: Island Press.

Nabhan, G. P., & Trimble, S. (1994). *The geography of childhood.* Boston: Beacon Press.

Piaget, J. (1969). *The child's conception of the world.* Totowa, NJ: Littlefield, Adams.

Pyle, R. M. (1993). *The thunder tree: Lessons from an urban wildland.* Boston: Houghton Mifflin.

Ratanapojnard, S. (2001). Community-oriented biodiversity environmental education: Its effects on knowledge, values, and behavior among rural fifth- and sixth-grade students in northeastern Thailand. Doctoral dissertation, School of Forestry and Environmental Studies, Yale University.

Savage, J. (1995). Systematics and the biodiversity crisis. *BioScience, 45*, 673–679.

Searles, H. F. (1959). *The nonhuman environment.* New York: International Universities Press.

Sebba, R. (1991). The landscapes of childhood: The reflections of childhood's environment in adult memories and in children's attitudes. *Environment and Behavior, 23*, 395–422.

Shepard, P. (1978). *Thinking animals: Animals and the development of human intelligence.* New York: Viking Press.

Shepard, P. (1996). *The others: How animals made us human.* Washington, DC: Island Press.

Sobel, D. (1993). *Children's special places: Exploring the role of forts, dens and bush houses in middle childhood.* Tucson: Zephyr Press.

Stegner, W. (1962). *Wolf willow.* New York: Viking.

Thomas, D. (1965). *Quite early one morning.* New York: New Directions.

Thomashow, M. (1995). *Ecological identity.* Cambridge, MA: MIT Press.

Untermeyer, L. (1949). *The poetry and prose of Walt Whitman.* New York: Simon and Schuster.

Weil, S. (1955). *The need for roots.* Boston: Beacon Press.

Wilcove, D. S., Rothstein, D., Dubow, J., Phillips, A., & Losos, E. (1998). Quantifying threats to imperiled species in the United States. *BioScience, 48*, 607–615.

Wilson, E. O. (1984). *Biophilia.* Cambridge, MA: Harvard University Press.

Wilson, E. O. (1992). *The diversity of life.* Cambridge, MA: Harvard University Press.

Wilson, E. O. (1993). Biophilia and the conservation ethic. In S. Kellert & E. O. Wilson (Eds.), *The biophilia hypothesis.* Washington, DC: Island Press.

6

Animals as Links toward Developing Caring Relationships with the Natural World

Olin Eugene Myers, Jr. and Carol D. Saunders

Animals are a compelling part of the human experience of the natural world. We can see evidence for this in the preponderance of animal images and metaphors in human mythology, folktales, art, creation stories, and other products of the mind, across many cultures (Urton, 1985; Lawrence, 1993). Indeed, one could argue that every segment of the natural world—plants, weather, landforms, waters, and so on—offers something surpassing and singular to the lives of people. But with animals, we believe we have a special case, a part of nature that is a potent and enduring part of our very development. Most important, we argue that animals provide a bridge to *caring* about the natural world in general.

One of the reasons animals are so fascinating to us is that they are highly responsive and offer many dynamic opportunities for interaction. We are social creatures, and animals appeal to our propensity to interact socially. As millions of pet owners and other animal fans can testify, animals provide intriguing and gratifying challenges that expand our abilities to understand them. Other parts of nature interact with us as well, but usually in more subtle and hard to pin-down ways. This elusiveness seems to have led many analysts to suggest that nature (including animals) matters to us only as we socially construct its symbolic meanings. Thus it can seem as if the meaning of nature boils down to nothing more than what our cultures make it out to be. This undercuts the ways that nature is compelling in its own right. But with animals, it is easy to see how nature gains significance through our direct experiences and immediate interactions with it.

Animals easily provide opportunities for such interactions because of fundamental aspects of our social development. As a child's social abilities develop, he or she perceives an animal as another being with subjective experience of its own. Once we accept the idea that animals may be social others to us with whom we can form relationships, other consequences follow. Specifically, animals are integral reference points for the child's sense of self, and thus they are very likely to be objects of human care. In exploring what this means we discover yet a further extension or at least a potential one. If you care about another—whether human or animal—you are likely to care about what that individual needs and the conditions that affect his or her well-being. This developmentally probable "natural care" about animals may lead to broader environmental caring.

The observation that we may come to care about nature or ecology by first caring about specific animals comes as no surprise, since today zoos, nature documentaries, and other agents of environmental education often build on this connection. However, the developmental bases of this potential have not been systematically explored. In this chapter we make a few steps in this direction, starting with animals as social others and then moving on to talk about care for animals and nature more broadly.

Early Social Interactions with Animals

Recently some favorite examples of child-animal interaction have been informal observations of the first author's 23-month-old daughter, Eva. She began to focus her gaze on animals at about age three months, first clearly registering our cats at age 12 weeks. By five months she was very interested in watching them move and in touching them (two weeks later she first noticed birds). Now Eva is drawn to many animals, from cows to craneflies. She approaches them and looks, often saying, "Hi, cranefly," for example. She's not entirely comfortable with that insect, but with a more interactive animal, like one of our cats, she attempts to instigate interactions—petting, brushing, and feeding. A typical bout includes approaching Java, reaching to pet him, or pursuing him to do so. If successful, she tries to hold and lift him or simply lays her head on him. She tells him, "I love you," and wants to see and touch his eyes and nose. She's learning he doesn't put up with that.

Eva and the cats work things out together, so her parents limit their facilitation but model positive behaviors. Nonetheless, she takes affection far beyond our behaviors, suggesting that her understanding of the cats arises in part from direct interaction. More evidence for this is how she acts toward the other cat, Mocha, who is cantankerous and snappy. By about nine months Eva was crawling and learning not to approach Mocha too closely. She did this even before she was warned. These days she often wants to pet him but obeys his snarls and keeps her distance (a feat of self-control). Her spontaneous relationships with these two cats are quite different.

Eva's interactions with Java and Mocha, while simple, resemble the actions of three- and four-year-olds with their classroom's guinea pig, Snowflake, as reported in Myers (1998). These episodes displayed several features, such as the child's overtures to establish an interaction, the excitement and emotion that children expressed when the guinea pig responded, and the mutual adjustments that made the interaction successful. For example:

> Mrs. Ray [the teacher] puts Snowflake on 31/2-year-old Rosa's lap, and gives her dandelion leaves to feed it. Rosa watches her, figures out how to present the leaves, and feeds her several. She cautiously offers the leaves, avoiding Snowflake's teeth. As long as the animal is eating, Rosa can pet her quivering body. Mrs. Ray returns and gives Rosa a blossom to feed her. Eventually she feeds it to the guinea pig; she looks over at me, wide-eyed with the fact Snowflake did eat it. (author's fieldnotes)

Events like these demonstrate how children deploy their social abilities in interaction with animals. To open the interchange, Rosa had to be able to hold still and offer food in a way the animal could perceive and respond to. And every micromove of feeding (how to hold the leaves, when to withdraw one's hand, and so on) required Rosa's responsiveness to Snowflake's moves. In return, the very possibility of this interaction depended on the guinea pig's having practiced and calibrated its moves with children before. Of course, Rosa was delighted when Snowflake unexpectedly consumed the flower.

Such observations of initiations, adjustments, and emotional reactions are not rare. The act of feeding an animal by hand seems to be especially powerful, and many early zoo memories revolve around this. But the details of these interactions have been bypassed by scholarly analysis.

Once we discern the structure of this dance of simple social moves, we will be able to appreciate the significance of animals in social development and the genuinely interactive bases of children's caring toward them.

Basic Domains of Social Relatedness with Animals

Animals are social others to children because of the cues children—and adults—read in interactions. These are the same basic cues we read in intraspecies interactions, and they fall into three domains or coherent sets, which we review here. These domains emerge sequentially as very early social and linguistic abilities emerge. Each is associated with an enduring sense of self in relation to others, as Stern (1985) argues, based on research with infants.

"Animate" Relatedness As shown by Stern (1985), by age three months and then through the life span, we encounter other people as *animate social others* because they display four invariants (or properties that remain constant across many different interactions)—agency, affectivity, coherence, and continuity. The same is true about animals (Myers, 1998). *Agency* is related to self-movement—for example, the guinea pig's motions in eating. *Affectivity* is most clearly shown in patterns and qualities of emotional arousal over time, such as Snowflake's excitement peaking in the first bite before settling in to busy munching. *Coherence* refers to bodily wholeness and congruity, conveyed in the coordinated movements of the animal on one's lap. And *continuity* is based on how repeated interactions can become regularized into a relationship, as we saw above with Rosa and Snowflake. These are the ingredients out of which more complex forms of interaction are built.

The domain of animate relatedness gives us the immediate and often bodily felt sense that a given animal or person might approach us, is relaxed, or moves and behaves with an injured, disorganized quality. We know what to expect of a specific individual we have been with repeatedly—and it of us. As a consequence, the animal appears to the child to have a living, subjective, individual, inner aspect; in short, it is a social other in a basic sense.

Just as do humans, animals display all the invariants mentioned above. But animals do so in ways that vary in degree from human patterns. Thus they offer "optimal discrepancy" as social partners for young children: they pose solvable challenges such as how to approach them so they don't run away, how to rouse them into action, and how to hold them. Furthermore, different animals require different accommodations. Children reveal that they register these gradient differences in interactiveness implicitly by actions that adjust to the animal's differences. They also reveal it explicitly in words or pretend play when they comment on or reproduce actions characteristic of another species. From these kinds of evidence we know that young children experience other animals as a differentiated community of social others. Shortly we discuss the consequences of this for the child's sense of self.

Animals and other people are the only entities that afford such experiences directly. Granted, it is possible to experience inanimate aspects of nature as subjective, but this appears to be more dependent on cultural construal, as in animistic religions. That animals are the key starting point for environmental caring is also suggested by findings that young children may not distinguish plants from inanimates (Carey, 1985; Margadant-van Arcken, 1989; Jaakkola, 1998; but see Inagaki & Hatano, 1996). Even after a biological domain of knowledge is developed, children apply psychological traits differentially to animals (Coley, 1995). Animals appear as living *and* feeling, making them potential objects of care.

Intersubjective Relatedness The next interactive domain to emerge developmentally is based on the ability to read nonverbal cues (from a care giver) that reveal that the infant's subjective state is understood and shared (a process called "attunement" by Stern, 1985). Probably very few interactions with animals are truly intersubjective in this sense. But animals do provide cues, which young children have learned from nonverbal and verbal interaction with care givers to read for shared attention, affect, and intentions. For instance, the excited barking of a dog in play could be read as indicating that the dog means to share the child's similar excitement with the play.

Such inferences are built partly on the child's emerging theory of mind and partly on the closely related pragmatics of verbal communication. The net result is that children sometimes read an animal's behavior in its context as indicating shared states such as mutual interest, liking, calmness, or other affects. For instance, Dawn, a girl in Myers's (1998) study, interpreted the turtle's crawling toward her as indicating it liked her. Five- to six-year-olds in the same study demonstrated a better grasp than did three- to four-year-olds of animals' limited capacities to share and confirm such states. The quasi-intersubjectivity that animals offer because they don't use nonverbal cues manipulatively (for example, to give mixed messages, redirect attention, or treat a topic as taboo) is not lost on children or adults. It creates a unique interactive realm between person and animal, enriching our analysis of their social interaction.

Linguistic Relatedness Language is the third domain to emerge, and it brings powerful new forces that interact with the other enduring domains. Clearly, animals share in this system of cues very incompletely at most. But just because young children *are* language users, they bring certain assumptions to their interactions with animals. Very young children experience language as just one part of a whole matrix of communication (Miller & Hoogstra, 1992). They readily assume animals take part in it, as revealed when they speak to them without modifying their speech from the ways they talk to other people. For example, a younger child in Mrs. Ray's classroom, Billy, yelled at the toad, as if it didn't respond because it couldn't hear. On the other hand, one older boy, Joe, developed a more differentiated theory—the idea that animals have their own humanlike languages that people may be able to understand. For language to work, certain assumptions must be made, and young children maintain these into their transactions with animals. These include that animals possess communicative goals and that animal action is meaningful and potentially decodable. Despite the putative "incorrectness" of these beliefs, the upshot is significant: looked at interactively, language does not set us apart from animals (its classical significance) so much as it puts our young on a track to wonder and care about them.[1]

As in the other domains of immediate social interaction, animals also provide something unique in relation to linguistic relatedness. Children creatively exploit animals' status outside the social forms that language is used to impose, such as proper behavior, ascribed roles, and precise diction. In animal-role pretend play (Myers, in press) and in their use of animals as confidants (Rochberg-Halton, 1985; Hoelscher & Garfat, 1993), children find a freedom from the demands for conventional word usage and syntax and from the accountability that language brings.

The Child's Self in Relation to Animals

The idea of self is often construed egocentrically, as if we really know only ourselves and merely make inferences about others by projecting onto them our own characteristics (even empathy is often conceptualized in this manner). And it is assumed that we also care about the self far more than about others. But the self-in-relation perspective holds that the self is not normally solipsistic nor so selfish. Stern's (1985) work shows that from the beginning the self is constituted through its *relations* with others who are distinct from the self; this is an important revision of centuries-old beliefs about infancy and the self in Western thought. Appealing to things we all have felt, Gendlin (1995) explained that it is not true that we sense others only in terms of our own experience. We can sense that we don't know what others feel, and they can also generate feelings in us that we have never had before (Gendlin, 1995). Gilligan and Wiggins (1987) describe this as "cofeeling," or the ability to participate in another's feelings and not just project our own feelings. The clarity of self and other may wane at times, and we can be confused about what is "my issue" and what is the other's. But development of the self is a dynamic process; gradually and simultaneously we come to know both self and other more completely. The self does not expand, but its *articulation* with vividly present others does.

While most attention by social psychologists has focused on our relationships to other humans, the dynamic process of clarification applies to our self in relation to animals as much as to other people. With animals the channels of relationship are slightly different, but young children's social abilities are developmentally plastic enough to include other species, with only partial distortion. The patterns of relatedness we have summarized above are robust and easily observed; children may be biologically and psychologically predisposed to exhibit them. Their consequence is that children differentiate the animal from the human, but not in a categorical fashion. Thus their sense of being a human self develops in relation to the available mixed-species community (those animals with whom the child interacts). The animal thereby becomes significant to the self *in its own right*, not primarily as a result of projective anthropomorphism. Just as with other humans, a child gets a sense of herself from how an animal acts back. We can infer that the meanings of animals to the child's self include a confirmation by a nonhuman of the self's own agency (notably, this can take positive forms or negative ones, as in animal abuse), coherence, affectivity, and continuity; clarification of what it means to be human; and a sense of connection across difference (Myers, 1998).

Relationships extend and clarify our awareness of similarities and differences between self and other, but they also expand the realm of our caring beyond our skin and beyond our psychological self. A naturally occurring consequence of self in relation is the propensity to take to heart the welfare of others to whom we are close. Here we have the durable psychological basis of an ethic of care toward animals and by extension toward nature more generally.

Caring Relationships with Animals

Qualities of Natural Care

To gain some insight into what caring for animals means to children, let's look at some examples from focus-group research we did at Brookfield Zoo (in Brookfield, Illinois). Over the last three years Brookfield Zoo has been conceiving and creating a new children's area and has sought to utilize research throughout the design process. The focus groups presented children of ages six, eight, 10, and 12 years with hypothetical situations and zoo plans and probed children's conceptions of superlative animal and zoo experiences. One scenario we asked them to imagine was a proposed exhibit area that offered the opportunity to help rehabilitate injured animals. It was unanimously popular. Why? In a session with 10-year-old boys, the following transpired:

Researcher: Maybe it [has] a broken arm or a broken leg. . . . Or maybe it lost its mother and it's a baby and it needs somebody to care for it. But it's an animal from the wild that needed some help. And you could—

Daryl: (interrupting) You can learn how to take care of something.

Randy: Yeah.

The idea of caring for animals' health had great appeal to the boys and was much more popular than cleaning cages as an activity. Randy explained why:

Randy: Because you can take care of the animals.

Daryl: Yeah, you can take care of something.

Researcher: Why aren't you taking care of the animal when you're cleaning the cage?

Daryl: You are taking care of the animal when you're cleaning out the cage, but—

Randy: It's just that's smelly.

Daryl: [Helping an animal directly is better] because you're being with a hurt animal and trying to help it.

Randy: Yeah. You helped an animal.

Researcher: But keeping its cage clean?

Daryl: Is somebody else's job.

There are several important points to note about the boys' conception of care. The first is that *care involves being open to the other's needs.* Note that what appealed to Daryl was the chance to "learn how to take care of something," implying that he would have to take in new information, including what this particular animal needed.

But these boys had a limited ability to be motivated solely by the other's needs and stopped short of wanting to clean cages. In contrast, consider the feelings of two eight-year-old girls, both of whom were attracted by the chance (again, in the scenarios we proposed) to care for animals, including the unpleasant aspects:

Researcher: So is [taking care of them] better than just getting to see them?

Veronica: Um-hm.

Researcher: Well, what makes that better? I mean, you have to go there and—

Karen: Because you can actually help—

Researcher: Okay. Some kids might say, I don't want to do that, that's extra work. Is that how you'd feel?

Karen: No.

Researcher: And why not?

Karen: Because it would be helping.

These girls illustrate a second quality of caring—*to truly put the other's needs first*.

For both groups, the care recipient's responsiveness is also an important factor. The animal can be seen to benefit from help, and its response to the care constitutes a form of reciprocity that makes caring satisfying. The *acknowledgment by the recipient* is a third quality of caring. Recall Daryl's emphasis on "being with" a hurt animal, which suggests he was attuned to how the animal receives the help.

Working in a relational framework similar to what we proposed above, Noddings (1984) analyzed care according to these same three qualities:

• Care involves receptivity, where the one doing the caring attempts to feel just what the other feels. This conception is consistent with our position above that the child's responsiveness to others is not based on empathy (where the self uses its own feelings as a model of the other's) or on the "expansion" of an essentially selfish self.[2] Both these conceptions presuppose a subject-object split. With Noddings (1984), Blum (1987), Gilligan and Wiggins (1987), and others, we emphasize cofeeling.

• When we fully receive the other, this catalyzes the second component of care: our motivations shift to give the other's needs and goals primacy over our own (Noddings, 1984). Granted, sometimes we "care" instrumentally in the service of self-interest, but a poor grasp of the other will result in inadequate care.

• The third quality identified by Noddings's analysis is acknowledgment of some sort from the one receiving the care. We perceive the ways animals act and flourish in response to our care as acknowledgment.

Caring of the sort we are discussing is an outgrowth of relationships and is often focused on individuals we know closely. Noddings (1984) calls this kind of caring "natural caring." In these cases, what "I want" is identical with what "I must do."

Morality Toward Animals

"Natural care," however, falls short of the impartiality of truly moral action since it may not guide us to override strong self-interest or to care for others with whom we have no personal connection. Nonetheless, it is probably an indispensable precursor.[3] What is of key importance is that the animal is perceived as the kind of thing whose subjective well-being can be harmed or benefited as a result of the self's actions.

Manifestations of early moral responsiveness to animals have been reported by Margadant–van Arcken (1984), Bailey (1987), Melson and Fogel (1988), and Poresky (1990). On the other hand, Kellert (Kellert & Westervelt, 1983; Kellert, 1985, 1996) found low moralistic attitudes in second-graders' responses to 54 questions tapping attitude categories. But other studies examining behavior in context and using more developmentally appropriate task demands suggest this conclusion may be premature. Myers (1998) found that instances of harm to animals were spontaneously—and urgently—reported by children to their parents, almost uniquely among many possible animal-related events in the preschool. For example, when a dead baby bird was brought to class, when a story was told about a puppy being accidentally stepped on, when the baby dove born in the classroom died, when the visiting spider monkey was confined to a cage: all these provoked concerned or even outraged responses from four- to six-year-olds.

In response to moral dilemmas entailing harm to both nature and animals, Kahn and Friedman (1995) and Kahn (1997) found that second-graders judged such actions wrong and that most children mentioned harm to animals in explaining their answers. Röver (1996, cited in Nevers, Gebhard & Billmann-Maheca, 1997) found that children as young as six years defended animals' interests on the basis of their being living things. The possibility that these are culturally biased findings is diminished by research extending Kahn's studies cross-culturally (Howe, Kahn & Friedman, 1996), where patterns similar those in the

United States were found for both urban and forest village Brazilian children. Moreover, even cultural relativists Shweder, Mahapatra, and Miller (1987) found that "kicking a harmless animal" was one of only a few acts that Oriya (Indian) and American five- to seven-year-olds agree is wrong.

The moral significance of animals should come as no surprise because in contrast to previously dominant views, young children's moral concern is now widely recognized as a highly probable developmental attainment.[4] The key developmental precursors—an appreciation of fairness and welfare, feeling with others, a sensitivity to standards, and (in situations of low to medium stress) an ability to overcome egocentrism and moral selfishness—now are well documented, with signs present by age two (Kagan, 1984; Packer, Theodorou & Yabrove, 1985). Kagan (1986, p. 88) observed, "Since W.W.I., American psychologists have declared you've got to teach children morality." Now we know it develops much more spontaneously. In fact, Kagan asserts that rather than needing parents to inculcate a moral sense, a child "can only lose a moral sense" (ibid.). Thus, instead of attributing moral failures to lack of instruction, we must look for factors that desensitize or set development back. Caring can be inhibited, for example, by the psychodynamics of distancing mechanisms.[5]

Early moral concern targets close individuals as the objects of care, but with development, the inadequacies of this approach gradually are revealed. The objects of care in some examples from the literature noted above pertain to animals that are distant from the self or even hypothetical. A bigger challenge occurs when caring about one animal precludes caring about another or about a human being. Caring obligations can be owed to multiple or conflicting recipients or can pertain to groups, systems, or even abstractions. Noddings (1984) admits that when the limits of "natural care" are reached, a moral sentiment of "I ought" must then compel the self to enter a relation of "ethical caring." In this way caring may be generalized beyond particular relationships. A parallel conclusion is offered by psychologists who have converged on the view that care and justice orientations are both essential in moral development. The justice orientation stresses impartiality and acting in ways that "I" might "not want to." While we may begin to care about individuals, situations of divergent goods or conflicts of principles call for justice-oriented moral reasoning to coordinate care among potential recipients.

A focus on ethical care is also compelled by our social interactional view of animals' place in human morality. Although animals are very real others to us in self-in-relation perspective, and although moral responsiveness and care toward them is probable, morality's origins and function in the human social group limit the likelihood we will extend it to animals. Human morality, tempering selfishness with altruism, is an adaptation to our highly interdependent and cooperative social environment. It is a strong force among people because our language lets us objectify, label, and evaluate behavior. Language thus allows us to create shared moral norms that work because those norms are confirmed *reciprocally* in social interactions between humans.

But animals generally do not demand, provide, withhold, or in other ways reciprocate moral consideration in return to our actions toward them. Thus, beyond natural care, their standing in the moral scheme depends on what we make it. In our culture, animals lack moral considerability equal to what we afford other humans, although the animal rights movement has changed this status in recent decades. With variations according to animal categories like pet, farm, pest, zoo, or wildlife, animals may be exploitable and expendable, tempered sometimes with humaneness. They cannot rise up as members of human liberation movements have, representing their own interests. Rather, if they have representation, it is voiced by humans to other humans. Too often, however, the emotional power of even trivial claims by other humans may trump urgent ones of animals (Partridge, 1996). And sometimes the opposite failure occurs—insistence on the absolute primacy of individual animals' rights over competing legitimate human or ecosystem needs. In reality, integrating obligations to animals, humans, and ecosystems is necessary but morally demanding.[6]

Environmental Care

Coordinating care for individuals with care for higher levels of biological organization clearly faces stiff psychological challenges, the full extent of which is beyond the limits of this chapter. Ecosystems are not unitary animate others that naturally recruit our moral emotional responses, and nature itself doesn't care in a human way about animals or interact with the self in a social way. Ecosystem integrity often conflicts with urgent human needs, and ecological concern sometimes faces similar denigration to that faced by concern for animals. Environmental care greatly increases the moral complexity of existence.

Even as we try to address moral concerns at the species level, we quickly realize there is a wide range of species to consider. Few would argue that there is no difference between mistreating plants and mistreating humans, but other distinctions may not be as obvious. We also realize that concern for individual animals or plants might lead to policies that are at odds with the best strategy for preserving species. Moral development occurs as one is confronted with questions such as whether to save an endangered species by exterminating individuals of an introduced species or by infringing on the life style of a traditional human culture. Unfortunately, we can't count on our natural care to guide such decisions. We might be very attracted to certain ecologically harmful species and not so attracted to organisms such as microbes and invertebrates that might nonetheless play important ecological roles.

According to Norton (1987), the reasons for protecting species can fall into different categories: *anthropocentric* reasons that locate the intrinsic value in humans, *biocentric* reasons that locate the intrinsic value in nonhuman entities, and reasons that highlight the *value of an experience* to fulfill an existing preference or alter such preferences. The last reason is related to the idea of transformative values. Experiencing natural objects can help adjust one's thoughts about one's place in a greater system. Species and ecosystems teach us about ecological relationships and provide analogies and metaphors that give us self-knowledge. Experiences of nature also provide opportunities for forming and criticizing our values. For example, a certain experience could promote questioning and rejection of overly materialistic and consumptive felt preferences. In all of these cases, species and ecosystems have a role, not because they have intrinsic rights but because they have an ability to form and transform values.

We believe that the formation of values and generalized care can grow from the strong caring for animals that children exhibit. Generalization entails conceptual growth, including the acquisition of knowledge of animals' habitat requirements, of interdependence within ecosystems, and of the ways humans affect environments. Given these inputs of knowledge, the extension of care from animals to habitats and ecosystems appears developmentally probable, building on the child's grasp of the animal as a subjective other with ecological needs. The process of consciously reconsidering preferences and consequently transforming values often begins with new experiences.

Role of Zoos in Facilitating Care About Animals

Direct interaction with animals is the starting point for *natural care*, and it often happens in families, at parks, on petting farms, and so forth. Zoos offer unusual opportunities to expand the child's mixed-species community beyond what can be kept at home or approached in the wild. The variety of experiences with a diversity of animals allows visitors to reflect on how they are similar to and different from other living creatures. Most people visit zoos as part of a social group, and together they create meaning from their encounters with animals. Memorable events are often related to animal activity or proximity (Saunders, Birjulin, Gieseke & Bacon, 2000). It appears that watching animals provides possibilities for self-in-relation experiences. The following are some examples of special moments reported by visitors at Brookfield Zoo:

> The lion was up at the window, so we were able to show it to our kids up close. Seeing it that close helps the kids get a true perspective on its size and awesomeness.
>
> How the gorillas were walking around and how they were eating, as if they were human.
>
> Dolphins seemed to stop at the underground viewing window to view us. It was wonderful to feel like you had somehow connected with these incredible creatures.
>
> I was mesmerized by the moon jellyfish. How can such delicate filmy things be living?

But observing animals is not as immediate and potent for the self as interacting with them. Young children in particular seem to enjoy the areas of the zoo where they can pet a goat, hold a guinea pig, or touch a snake. While zoos can provide only limited opportunities for tactile contact because of the large number of visitors and potential stress for the animals, they can model caring behavior for parents. As a result, parents can gain the confidence to provide similar animal experiences for their children at home and perhaps prevent some of the obstacles to care. So long as such experiences build on first-person relationships with more familiar animals, the connection can be made vicariously or through how the experience is framed. Even connecting to remote, nonanthropomorphic, microscopic, ugly, inert, or other "nonsocial" animals is possible but may require encapsulating them in narratives.

In addition to helping to inspire natural care, zoos facilitate *environmental care* and moral concern by sharing stories about caring for animals at different levels. Most obviously they care for the well-being of individual animals and can thus be role models. The difficult choices about animal welfare at zoos are carried out against the background of sometimes desperate species-survival situations in the wild. Zoos manage animals at the species level through cooperative breeding programs with other zoos. Knowledge gained from animals in captivity has helped manage animals in the wild and vice versa. For some rare species, zoos are their last chance before extinction. But zoos can't save all the animals in the world that need help. Even as zoos become increasingly involved in *in situ* conservation, there is still limited space. Thus, although zoos care for their animals at the individual, species, and ecosystem levels, their biggest impact may be in the ways they encourage their visitors to do the same. The more naturalistic exhibits of modern zoos allow many possibilities for stories of animals in the context of their habitats and ecosystems. Featuring local animals can help celebrate regional environments and sense of place. Zoos are also playing a larger role in suggesting things that people can do in their everyday lives to help conserve biodiversity.

Developing Caring toward Nature

Despite the moral complexity that lurks in our relations with animals, there is a strong case, as we have seen, that caring for them arises reliably in early child development. Furthermore, this caring extends in some obvious and simple ways to caring about habitats, species, and ecosystems. Let us return to some fresh examples to contemplate the potentials of extending generalized care beyond individual animals. In another focus group at Brookfield Zoo, 12-year-old Mary, Marcy, and Tess were just explaining they did or would like to chase frogs. Sheila, however, led the conversation in a new direction:

Sheila: I'd like to see if the water is polluted and see if there's any dirt on him and see how they feel and stuff.
Researcher: How come you're checking that out—to see if the water is polluted or if there's dirt on him?
Sheila: If it's polluted, it's kind of like nasty. It's like animals shouldn't be living in polluted water or polluted air and stuff like that. If it's out somewhere nice, like a farm or something, I wouldn't think they should allow pollution up there.
Researcher: What would you do if it was—if the frog looked like it was polluted? What would you do?
Tess: Try to help it. I probably would take it to an animal shelter that accepted different kinds of animals other than dogs and cats. Or bring it to a zoo or something because they have frogs and stuff.
Mary: See if they could possibly clean the river or clean whatever their home was, maybe clean them up.

The girls went on to explain how they might "take a stand" or raise money to help. Marcy told about an actual incident when she had taken action:

Marcy: Me and my friend, when we went [to a place near a farm] once, there was a whole bunch of dead frogs on the side of the river and fish. There was a whole bunch of garbage, sort of damming up the river, and a lot of fish and frogs and other animals were dead. So we started cleaning it up. I was staying there for a week, so we cleaned it up. We went out there every day, and we cleaned it up and actually got all the garbage out. I guess it's okay now, but I haven't been there.

For these girls, it was concern about animals that moved them to care about nature more broadly. How does concern with the welfare of individual nonhumans expand to wider systems and still include consideration for individuals? One likely extension is in terms of how the animal's needs are perceived, as a 10-year-old explained:

Researcher: How about you guys? Do you feel like it's important to try to understand animals? Is that something you want to do? Yeah? Tell me about that, Linda.

Linda: Well, it might be nice to understand the animals because you'd be able to know what they need and what they don't need. And what they don't like.

Note that in Linda's view the subjective "likes" of animals are linked to their "needs"; both terms reference the animal's experienced well-being. A feeling for the animal's inner life is the thread that guides children into caring about its world.

What do they find when children take such needs into consideration? A study on "What do children think animals need?" at Brookfield Zoo is asking this question (Myers, Saunders & Garrett, in preparation). One answer is habitat. Thirteen-year-old Michele showed what a lion needs:

Researcher: So I'm wondering if you can take some of the markers and draw in everything that you think that he needs.
Michele: OK. He like hides in the grass here.
Researcher: And why do you think he might be hiding?
Michele: Well, he needs to hide so that the animals that he hunts don't see him.

Supporting the extension to habitat, Kahn (1999) found that welfare of animal residents was a reason frequently given by children across three cultures for judging the pollution of a waterway as morally wrong. Indeed, as with caring for individual animals, the human self is still very much in the picture in caring for animals in their habitats. In the same study, 11-year-old Alan described efforts to save elephants:

Alan: We need like maybe people protecting them. Like maybe putting them in zoos. Or like once in a while somebody would like take a population, like a count, of them. And then if they come back in maybe like a year or a month or something, to see how many have disappeared. And if there's some disappearing, they might take more to zoos. Or maybe people might bring them somewhere else to save them.

Alan expands caring by multiplying the number of individuals cared for, while acknowledging both positive and negative impacts by humans on habitat needs. Although neither of these examples is very ecologically sophisticated, with additional knowledge they could be.

We can gain a glimpse of how early experience with animals continues to animate care for nature from two participants in a study of young adults' environmental career choices (Myers, in preparation). Here we find a maturing capacity for practical caring for nature. Shari, age 20 and a college junior majoring in environmental studies, retrospectively connected her earlier childhood animal interests with her career desire to make a difference:

> When I was younger, I used to raise frogs and save all the frogs and amphibious creatures. I was really an outdoors kid. And when I saw actually how houses moved in and everything seemed to die off, when I hit about ten, and so that kind of disturbed me. And everywhere everything was disappearing, and I was like, "What?" Maybe all the ponds and the drainage systems were changing in the area because of the houses. So I decided that I wanted to do something to preserve estuary wetland areas. It was mainly like, "Oh, I'd like to save them." . . . [Now] I'd like to contribute. It's such an intricate subject that we know so very little about it as a whole, since there's just thousands upon thousands of undiscovered species. And there's plenty of things to do. I just figured that I could find my niche very easily in that area, especially when I started reading about frogs being indicator species, absorbing the UV rays. . . . I did find something that I was really passionate about, really simply.

Striving now to grapple with complex and nonimmediate human-environmental interactions, Shari is still motivated by frogs, across large developmental strides. There is a vast range of careers and avocations that embody concern for animals and nature and that require conceptual and practical balancing of respect and care for different levels of organization in nature. The practical activity of caring for animals, species, and ecosystems is best conceived of not simply as physical activity but more fundamentally as moral activity. Another college junior, 21-year-old Scott, expressed this well:

> And when I'm working on things that I think really matter, like composing an article for *The Planet* [an award-winning student environmental advocacy magazine], and I feel a sense of duty, and I feel a sense of direction, and I feel like I'm actually achieving something.

Actions expressing caring for nature embody values. They are performances, often undertaken with others, that inspire self and others in society, sometimes through passionate and personal debate and other times through quiet example. Other college students anticipated activism, education, doing community outreach, or expanding environmental discourse to include social justice. Although the early precursors of caring are developmentally prepotent, these mature expressions depend on—but also create and critique—a supportive cultural context for their flowering.

Looking much further out the developmental spectrum, connection and concern for the well-being of individual animals and nature more generally is something that carries through adulthood for many. An exemplar in this regard is Lois Garlick, a long-time member of the North Cascades Audubon Society, now in her eighties. For many years Lois has cared for injured birds in her backyard animal hospital and aviary. Some come to stay, while others recover and are released. Lois doesn't reject even birds like young starlings that others might spurn. But what makes Lois remarkable is that she embraces individuals at the same time as she defends larger ecological systems. She and her husband, George, have long been caretakers for a Nature Conservancy island. For years she has been a watchdog on shoreline management and other conservation issues. Informed about ecology (including starlings' invasive pattern) and having mastered small and large administrative, legal, and political processes, she reads development notices in the local paper, investigates, and takes action. She has inspired and nurtured a second generation of local environmental activists, at the same time as she networks with other local wildlife rehabilitators, helping create the very conditions of caring. Lois's relations to nature exemplify the balancing of care for several levels of organization in a world that offers humans a home—a home for which she also takes her share of responsibility in running.

Conclusion

In exploring the potential that animals offer for an expansion of caring to nature, we have taken a *social interactional* perspective. We have sought to ground a theory of relations to nature in knowledge of developing human social abilities. We have argued that human social proclivities lead children to respond powerfully and flexibly to individual animals. The social responsiveness children show to animals has close links with cognitive, emotional, and moral development. And caring for animals in these ways extends with development beyond animals to species, ecosystems, and nature broadly. Direct interaction with animals, which is the starting point for natural care, can occur in many places, such as families, parks, and zoos. Zoos can also facilitate environmental care and moral concern by sharing stories about caring for animals at different levels.

Caring about animals and nature is related to the idea of biophilia. We agree with Kahn (1999) that an adequate conception of biophilia must be developmental because the person negotiates the intervention of environment between genotype and phenotype. Further, rather than viewing biophilia as a result of specific adaptive preferences for biotic environmental features, we would view it as in part a by-product of our species' social evolution. Traits that are thought to be distinctive about our species—such as our docility, cognition, language, and morality—arose in response to selection in a complex ecological but also social environment. As Humphrey (1984) argued, the leap in complexity of social interaction as hominids evolved to *Homo* called forth a leap in the psychological understanding of self and others. This leap involved extensive decoupling of automatic responses from their stimuli, such that the social responses of our species are not tightly targeted to only conspecifics. Our argument has been that social development imparts universal roots and dynamics to relations to nature, which thus contributes uniquely and irreducibly to our development. Understanding how children care about animals and nature involves determining how children's social development constrains and enables an unfolding sense of self in relation.

In sum, our chapter presents an account of biophilia grounded in our species' social development. Caring about nature broadly may begin by caring about individual animals. This early care is developmentally probable given some reliably occurring conditions—the child's propensity to grasp animals as animate social others and normal early moral development (including the absence of obstacles). Care about individual animals develops "naturally" out of relationships. It involves being open to the other's needs, truly putting the other's needs first, and perceiving the other's response to care. When children also have information about ecological dependencies, they discover that caring about animals means caring about habitat and ecosystems. This environmental caring involves the formation and transformation of values and a type of caring that is generalized beyond particular relationships. Challenges to environmental caring include the need to balance different recipients of care, to resolve conflicts, and to consider the health of whole systems. But these challenges do not exclude care for individuals and bear a formal resemblance to those of intrahuman moral development. When care eventuates in taking responsibility for human action in nature, it offers lifelong opportunities for connection and making a difference.

Notes

1. Yet language is a double-edged sword: while it allows the discovery of commonality with animals and metaphorical similarities broadly, it also allows us to reify meanings, mistaking the label for the thing. Thus language makes it possible for us to believe that the difference between humans and animals is categorical rather than one of degree. But taken alone as a basis for a theory of human-nature interaction, this latter potential of linguistic categories produces an incomplete analysis (Myers, 1999).

2. Plumwood (1991) has articulated the logical and psychological deficiencies of several ways of construing an ecological self.

3. In joining emotion and morality, we are in agreement with Schopenhauer (1841/1965), who held that compassion underlies the capacity for moral motivation, and with Hume (1777/1975), who declared "sympathy" to be a key and universal *moral* "sentiment" and determiner of other-oriented action (see also Kagan, 1984). Even Kant, whose moral system placed reason at its center, admitted that antecedent "feelings" including "love" (in the sense of charity) lie "at the basis of morality" (Beehler, 1978, p. 128, quoting Kant, 1797/1964, p. 59). Developmentally, Hoffman (2000) has elaborated many ways that empathy contributes to morality. We denote the emotion that draws us beyond ourselves as caring.

4. Robust moral sensitivity may be underlain by the generalization of universal early experiences, particularly experiences of relative powerlessness and vulnerability to abandonment, as suggested by Gilligan and Wiggins (1987).

5. Ross (1983, p. 204) has argued that "to perceive another's distress as distress is to perceive it as *prima facie* a bad thing . . . [and] to be avoided or prevented." Ross suggests that altruism is more than a mere possibility for our species. When response to perception of another's distress is absent, we may assume it is blocked by some other process. These processes include particularly the distinctions we make between deserving and undeserving or between objects of concern and of indifference. If such distinctions are absent, another's distress is likely to motivate us to act. Serpell (1986), Plous (1993), and others have theorized that tolerance of harm to animals requires the adoption of psychological distancing mechanisms such as misrepresentation of the harm, shifting the blame, denial,

rationalizing, deindividuation, or otherwise reducing the cognitive and emotional conflict. One defense, out-grouping the victim, is highly elaborated in our cultural discourses that explain and rationalize humans as superior to animals. Such factors reify differences in degree into morally loaded categorical divides. When such mechanisms are ambient, individuals may adopt them, thereby minimizing intrapsychic conflict but also compromising the capacity to care.

6. An account of morality adequate to the cognitive, emotional, psychodynamic, and interpersonal dimensions of moral functioning that characterize our moral relations with animals is offered by Haan, Aerts, Cooper (1985) and Haan (1991).

References

Bailey, C. (1987). Exposure of preschool children to companion animals: Impact on role-taking skills. *Dissertation Abstracts International, 48*(8), 1976A.

Beehler, R. (1978). *Moral life*. Towota, NJ: Rowman & Littlefield.

Blum, L. (1987). Particularity and responsiveness. In J. Kagan & S. Lamb (Eds.), *The emergence of morality in young children* (pp. 306–337). Chicago: University of Chicago.

Carey, S. (1985). *Conceptual change in childhood*. Cambridge, MA: MIT Press.

Coley, J. (1995). Emerging differentiation of folkbiology and folkpsychology: Attributions of biological and psychological properties to living things. *Child Development, 66*, 1856–1874.

Gendlin, E. T. (1995). Crossing and dipping: Some terms for approaching the interface between natural understanding and logical formulation. *Mind and Machines, 5*, 547–560.

Gilligan, C., & Wiggins, G. (1987). The origins of morality in early childhood relationships. In J. Kagan & S. Lamb (Eds.), *The emergence of morality in young children* (pp. 277–305). Chicago: University of Chicago.

Haan, N. (1991). Moral development and action from a social constructivist perspective. In W. M. Kurtines & J. L. Gewirtz (Eds.), *Handbook of moral behavior and development*, Vol. 1, *Theory* (pp. 251–273). Hillsdale, NJ: Erlbaum.

Haan, N., Aerts, E., & Cooper, B. (1985). *On moral grounds: The search for a practical morality.* New York: New York University Press.

Hoelscher, K., & Garfat, T. (1993). Talking to the animal. *Journal of Child and Youth Care, 9*(3), 87–93.

Hoffman, M. (2000). *Empathy and moral development: Implications for caring and justice.* New York: Cambridge University Press.

Howe, D., Kahn, P. H., Jr., & Friedman, B. (1996). Along the Rio Negro: Brazilian children's environmental views and values. *Developmental Psychology, 32*, 979–987.

Hume, D. (1975). *Enquiries concerning human understanding and concerning the principles of morals.* Oxford: Clarendon Press. (Original work published 1777)

Humphrey, N. K. (1984). *Consciousness regained: Chapters in the development of mind.* Oxford: Oxford University Press.

Inagaki, K., & Hatano, G. (1996). Young children's recognition of commonalities between animals and plants. *Child Development, 67,* 2823–2840.

Jaakkola, R. O. (1998). The development of scientific understanding: Children's construction of their first biological theory. *Dissertation Abstracts International: Section B,* Vol. 59(6-B), 3094.

Kagan, J. (1984). *The nature of the child.* New York: Basic Books.

Kagan, J. (1986). Presuppositions in developmental inquiry. In L. Cirillo & S. Wapner (Eds.), *Value presuppositions in theories of human development* (pp. 63–88). Hillsdale, NJ: Erlbaum.

Kahn, P. H., Jr. (1997). Developmental psychology and the biophilia hypothesis: Children's affiliation with nature. *Developmental Review, 17*(1), 1–61.

Kahn, P. H., Jr. (1999). *The human relationship with nature: Development and culture.* Cambridge, MA: MIT Press.

Kahn, P. H., Jr., & Friedman, B. (1995). Environmental views and values of children in an inner-city black community. *Child Development, 66*(5), 1403–1417.

Kant, I. (1964). *The doctrine of virtue* (H. J. Paton, Trans.). New York: Harper & Row. (Original work published 1797)

Kellert, S. R. (1985). Attitudes toward animals: Age-related development among children. *Journal of Environmental Education, 16*(3), 29–39.

Kellert, S. R. (1996). *The value of life.* Washington, DC: Island Press.

Kellert, S. R., & Westervelt, M. O. (1983). *Children's attitudes, knowledge and behaviors toward animals (Phase 5).* Washington, DC: U.S. Fish and Wildlife Service.

Lawrence, E. A. (1993). The sacred bee, the filthy pig, and the bat out of hell: Animal symbolism as cognitive biophilia. In S. R. Kellert & E. O. Wilson (Eds.), *The biophilia hypothesis* (pp. 301–341). Washington, DC: Island Press.

Margadant–van Arcken, M. (1984). "There's a real dog in the classroom?" The relationship between young children and animals. *Children's Environment Quarterly, 1*(3), 13–16.

Margadant–van Arcken, M. (1989). Environmental education, children and animals. *Anthrozoös, 3*(1), 14–19.

Melson, G. F., & Fogel, A. (1988). Children's ideas about animal young and their care: A reassessment of gender differences in the development of nurturance. *Anthrozoös, 2*(4), 265–273.

Miller, P. J., & Hoogstra, L. (1992). Language as tool in the socialization and apprehension of cultural meanings. In T. Schwartz, G. White & C. Lutz (Eds.),

New directions in psychological anthropology (pp. 83–101). Cambridge: Cambridge University Press.

Myers, O. E., Jr. (1999). Human development as transcendence of the animal body and the child-animal association in psychological thought. *Society and Animals, 7*(1), 121–140.

Myers, O. E., Jr. (1998). *Children and animals: Social development and our connections to other species.* Boulder, CO: Westview Press.

Myers, O. E., Jr. (in preparation). Environmental identity and college-age development.

Myers, O. E., Jr. (in press). Young children's animal-role pretend play. In R. Mitchell (Ed.), *Pretending in animals and humans.* Cambridge: Cambridge University Press.

Myers, O. E., Jr., Saunders, C., & Garrett, E. (in preparation). What do children think animals need? Ecological, Aesthetic and psycho-social frame works.

Nevers, P., Gebhard, U., & Billmann-Maheca, E. (1997). Patterns of reasoning exhibited by children and adolescents in response to moral dilemmas involving plants, animals and ecosystems. *Journal of Moral Education, 26*(2), 169–186.

Noddings, N. (1984). *Caring.* Berkeley: University of California Press.

Norton, B. G. (1987). *Why preserve natural variety?* Princeton, NJ: Princeton University Press.

Packer, M., Haan, N., Theodorou, P., & Yabrove, G. (1985). Moral action of four-year-olds. In N. Haan, E. Aerts & B. Cooper (Eds.), *On moral grounds: The search for a practical morality* (pp. 276–305). New York: New York University Press.

Partridge, E. (1996). Ecological morality and nonmoral sentiments. *Environmental Ethics, 18*(Summer), 149–163.

Plous, S. (1993). Psychological mechanisms in the human use of animals. *Journal of Social Issues, 49*(1), 11–52.

Plumwood, V. (1991). Nature, self and gender: Feminism, environmental philosophy and the critique of rationalism. *Hypatia, 6*(1), 3–22.

Poresky, R. H. (1990). The young children's empathy measure: Reliability, validity and effects of companion animal bonding. *Psychological Reports, 66,* 931–936.

Rochberg-Halton, E. (1985). Life in the treehouse: Pet therapy as family metaphor and self-dialogue. In M. Sussman (Ed.), *Pets and the family* (pp. 175–189). New York: Haworth Press.

Ross, A. (1983). The status of altruism. *Mind (N.S.), 92*(366), 204–218.

Röver, M. (1996). Die Entwicklung kinderlicher Einstellungen gegenüber Tieren. Ergebnisse von Gruppendisckussionen in der 3. und 9. Klassenstufe. Master's thesis, University of Hamburg, Hamburg, Germany.

Saunders, C. D., Birjulin, A. A., Gieseke, T. J., & Bacon, L. (2000). Visitor satisfaction at Brookfield Zoo. Manuscript, Communications Research, Brookfield Zoo, Brookfield, IL.

Schopenhauer, A. (1965). *On the basis of morality* (E. F. J. Payne, Trans.). Indianapolis: Bobbs-Merrill. (Original work published 1841)

Serpell, J. A. (1986). *In the company of animals: A study of human-animal relationships.* New York: Basil Blackwell.

Shweder, R. A., Mahapatra, M., & Miller, J. G. (1987). Culture and moral development. In J. Kagan & S. Lamb (Eds.), *The emergence of morality in young children* (pp. 1–83). Chicago: University of Chicago.

Stern, D. (1985). *The interpersonal world of the infant.* New York: Basic Books.

Urton, G. (Ed.) (1985). *Animal myths and metaphors in South America.* Salt Lake City: University of Utah.

7

Animals in Therapeutic Education: Guides into the Liminal State

Aaron Katcher

This chapter seeks to explain the results of interaction between children and animals in therapeutic settings. In the presence of animals, highly aggressive children behave more cooperatively, become less antagonistic, and display greater social competence (Kaye, 1984; Ross et al., 1984; Katcher & Wilkins, 1993, 2000). Children with a wide variety of diagnoses who are nonverbal or withdrawn in the presence of adults become more socially interactive and are able to engage in dialogue more freely with less physiological arousal (Fine, 2000). The findings suggest that the presence of the animal permits a favorable redefinition of both the self and adult care givers. This alteration in social attractiveness is similar in many respects to that seen when normal adults interact in the presence of pets (Friedmann, 2000). While the biophilia hypothesis might explain why animals hold children's attention and lower arousal, and ideas centered around neotony of domestic animals might predict the decrease in aggression, there is no good explanation of the favorable alteration of social perceptions. This chapter explores an explanation of those effects in terms of two ideas: Victor Turner's idea of liminality from the domain of anthropology (Turner, 1982) and D. W. Winnicott's idea of the transitional object from psychoanalysis (Winnicott, 1971).

Animals, Children, and Therapy

For the past 10 years Dr. Wilkins and I have been studying the effects of therapeutic education structured around care of animals and nature study in the residential treatment of children with autism, developmental disorders, attention-deficit hyperactivity disorder (ADHD), conduct disorder, and oppositional-defiant disorder. We have also observed the effects of similar programs in three public schools with children in special education who are usually diagnosed as seriously emotionally disturbed (SED) or learning disabled (LD). Children have been studied in nine settings, and the results have been documented with a controlled clinical trial at one residential treatment center, correlational studies at the same center, clinical case histories, and analysis of critical incidents (anecdotal information) at all nine centers.

The model for the treatment method was developed for residential treatment of severe ADHD, oppositional-defiant disorder, and conduct disorder and then later applied to children with other diagnoses. Although the treatment methods and results have been reported elsewhere (Katcher & Wilkins, 1993, 1998, 2000), it is necessary to describe them briefly to justify the theoretical analysis offered below.

In all of the nine different sites the center of activity was a building or a classroom housing a collection of animals that was referred to as a "companionable zoo" and that was administered by procedures outlined in a manual of operations (Katcher & Wilkins, 2001). The animals common to all the companionable zoos included rabbits, gerbils, guinea pigs, birds (most often parakeets, cockatiels, and parrots), ferrets, chinchillas, iguanas, turtles, frogs, and tropical fish. In six facilities there was also a barn and an outdoor paddock area for small farm animals. These included miniature horses, dwarf goats, pot-bellied pigs, sheep, and chickens. Some zoos had more exotic animals such as wallabies or sugar gliders. It is our impression that the kinds of animals were not significant, although their number may have been, in that the zoos needed enough animals to create an environment that was distinctly different from an ordinary classroom. It was also important that most students were unfamiliar with at least some of the animals as fear and novelty played roles in directing and shaping the children's responses. In addition to the cages, tubs, and tanks for the animals, the rooms contained space for storage of feed and bedding, a table for conferences and discussions, a small library of nature magazines and field guides, and computers for working with the many compact discs devoted to animals, ecology, and natural history. There was sufficient wall and shelf space for the display of artwork and "found objects" from nature walks. In the classrooms there was a collection of plants in the windows, and the zoo buildings had gardens nearby.

In the six residential settings and the private day school the children came to the on-site companionable zoos for two to five hours a week during regular school hours, during which time specially trained instructors directed their activities. In the public schools, the animals were housed in the special education classrooms, but instruction centered on the animals or interaction with the animals also occupied about two to five hours a week. The regular special education teacher was responsible for the zoo program.

On starting the zoo program the children first learned a general moral orientation toward the animals and natural settings. They were told that the animals' welfare is a central concern that takes precedence over teaching or recreational activities. They were also given two prime directives: talk softly and move gently around the animals, and respect the animals, other students, and staff. Having students sign contracts reinforced the overarching moral structure of the zoo program. One contract commited students to general care of all of the animals, another more specific contract was signed when a child adopted an animal, and a third was signed when a student wished to breed an animal and pledged to find a place for any offspring. The zoo was the only place in the institution that had a purpose and moral climate that was not entirely focused on the individual student. In the zoo the animals had priority over other concerns, and the child was expected to do more than not break rules. He or she was expected to participate in the work of animal care and actively support, through cooperation with peers and staff, the welfare of the animals. The students responded to this demand by perceiving the work of the zoo to be real and meaningful.

Elsewhere in the institution the children's behavior was controlled by a point system. Such systems give students points for good behaviors (as defined by the institution) and subtract points for bad behaviors. The points can be exchanged for goods or privileges. The systems are usually heavily weighted toward the avoidance of bad behaviors so that time without antisocial behavior earns the most points. Helping other students with homework or chores usually does not earn points. In general, cooperative or prosocial behavior is poorly rewarded. As a result the student is focused almost exclusively on earning points for his own benefit and has no responsibilities for the welfare of his peers. His behavior is valuable only when it earns him rewards. Morality is reduced to economic self-interest. The students recognized that the reward system was arbitrary and had no meaning outside of the institution.

Learning in the zoo programs was broken up into small units. Most units centered on an animal in the zoo and contained both a set of skills to be mastered (cleaning a cage, changing food and water, or holding an animal properly) and a knowledge set that contained some general statements about its biology and behavior and explained the requirements for the animal's care. Other units explored activities such as gardening, visiting state parks, visiting pet stores, camping, fishing, fire safety, identifying poisonous plants and insects, and gathering food in the wild. These units activities combined skills, factual information, and moral instruction.

The moral instruction was offered in association with the tasks performed at the zoo or the demands of the places being visited. The units of knowledge described proper social behavior in pet stores, at state farm expositions, in state and national parks, on camping expeditions, and at other schools or institutions during demonstrations of their animals. In the zoo children were frequently reminded of the prime injunctions to behave gently and respect others. Where problems with taking turns or putative insults arose, they were talked through their outrage and asked to consider the other's point of view.

During the period of observation zoo animals became the reference point for moral reasoning. Students were asked to think about their animals' needs, wants, rights, and state of mind as a preliminary step for reasoning about the condition of other people. The child's social perspective was first directed toward the animal, which could be reasoned about because the child did not deploy a set of automatic negative defensive attributions to explain the animal's behaviors. For example, children did not become angry when bitten, as they were with some frequency, by the small rodents in the zoo's collection. They explained the biting as defensive: "He was frightened," "I held him too tightly," "I reached in the cage too quickly." Nor did they interpret the animal's defensive activity as dislike. If an animal responded to their attempts to hold it by struggling or actively attempting to escape, its behavior was legitimized in terms of some need of the animal or error in approach by the child. One child who was permitted to adopt an irascible adult chinchilla persisted in trying to gentle the animal for two months before giving up, saying tearfully, "It isn't that I don't like him. I think he's just mean." While "being mean" was immediately offered as an explanation for human behavior, it required several months of testing before this boy was willing to attribute it to an animal. The concept of respect was difficult for the children to grasp in a purely human context, where for them it denoted only the recognition of dominance, but they could apprehend it when applied to the animals because no salient competition existed between child and animal. Thus, after establishing some consensus with the instructors about what the "legitimate" wants, needs, and motivations of animals entailed, the children could then begin to reason about people. This moral reasoning that moved from contemplating the animal to developing new ideas about people was facilitated by their play, in which the animals took on a variety of different identities and played out a set of human roles. The imaginary society of animals playing human roles was always more benign than their own working model of human society. The roles that the animals played as bearers of human attributes were noted by the school psychologists. They observed that events in the zoo like births and deaths were talked about in therapy sessions and that the discussion then moved to previously repressed reactions to similar events in the child's family life.

The use of animals as characters in morality plays was part of the instruction, but the children spent as much or more time playing with their animals. That play could be pure fantasy, with the animals as characters and with piping and scraps of wood as props to give fanciful purpose and destination to their scurrying. When rain turned a gravel walk into an ephemeral pond, the children brought the turtles and iguanas outside and played *Jurassic Park*. They drew pictures of their animals in various roles and guises that could be ranked along an axis ranging from realistic to anthropomorphic to monstrous. They wrote stories about their pets that almost always linked the author and the pet in joint action or shared identity. Their play combined what they learned from their lessons and experience about the animal's reality with all the attributes of animals in folk tales, films, cartoons, and fairy tales.

Although the quantitative data have been gathered largely from children with attention deficit disorder and conduct disorder, the results of enlistment in the activities of the zoo have been consistent across all diagnostic categories. Contact with animals and natural settings was an effective means of entraining and holding the children's attention. The direction of attention was also associated in inhibition of rapid physical responses (a problem in these children), especially when the situation was novel or the children were anxious or uncertain. The inhibition of physical responding in turn was associated with increasing time for reflection and more verbal behavior in the form of questions about the animal. Those questions generated a teaching dialogue where the knowledge offered matched the children's need to know. The teacher, the animal, and the student created a zone of proximal development (in Vygotsky's terminology, 1986), where the social phase of learning could take place (Tharp & Gallimore, 1990). The capacity of animals or nature to focus attention extended to the lessons formally structured around animals. Even children with quite limited intelligence or capacity to follow verbal directions persisted in learning the skills and information necessary for them to handle the animals.

When the children entered the program, there was an immediate decrease in hostile and aggressive behavior. In the campus where the controlled study took place, fights or aggressive episodes requiring physical restraint of students were daily occurrences. In the first six months of the controlled study we would have expected 35 physical restraints during the time the children were in the zoo program, but we observed none. In the nine years since the conclusion of the study it has never been necessary to restrain a child in any of the residential programs. Any problems with aggression that occurred were managed by brief periods of "time out" from zoo activities. Since there have been 11 instructors over the years in the zoos, and none of them have had to use physical restraint, we are confident that the decrease in aggression is a true effect of the program and not a result of the extraordinary personal skills of a few teachers.

Cooperative behavior between peers was much more obvious in the zoo program than anywhere in the rest of the institutions. Children helped each other with cleaning cages, the zoo, or the grounds. They worked with the zoo instructors to gather and stack firewood, which was a means of earning money for the zoo. They helped each other in a variety of ways on camping, fishing, and hiking trips. Angry taunting provocative behavior was not absent by any means, but it was subdued, could be controlled by verbal intervention from the instructors, and did not escalate.

The change in behavior toward peers was dramatic, but the children's response to adults in the zoo was perhaps the most distinctive aspect of the program's ambiance. The children's relationships with the zoo instructors had qualities not seen in the rest of the institution. The children accepted the authority of the zoo instructors as legitimate and not imposed by force or institutional control. They approached them for information, accepted their decisions, and gave them the status of experts in domains that were both important to them and part of what they saw as the real world. They wanted to be in physical proximity to the instructors—in contrast to their pattern of avoiding most adults in the rest of the institution. They were also more adept in their style of interaction and displayed more social skills than they apparently possessed in other circumstances. These social skills were readily deployed when visitors came to the zoo. The children greeted them, offered to show them around, answered questions, brought them animals to hold, and asked them questions about their identity and the purpose of the visit. The same aplomb was shown when the children demonstrated their animals to other classes of children, to adults in a senior center, to patients in a closed head injury ward, and to groups of teachers, social workers, and therapists. Their behavior and the skill with which they handled the animals made the visitors or their audience feel comfortable and well disposed to the children. The most frequently asked question was "Why are these children in residential treatment?"

The favorably benign impression of the visitors was also mirrored in the students reevaluation of themselves. If we contrasted measured self-esteem using the Piers-Harris Children's Self-Concept Scale, we found that their self-impression was significantly more favorable in the zoo than in the classroom. This index is designed to be a measure of persistent change, but we found that the zoo environment had a powerful contextual effect on self-concept.

All these behavior changes were initially limited to the confines of the program or to the times when the children were in contact with the animals or natural centers in the presence of the zoo instructors. Cooperative and friendly behavior could dissipate immediately as soon as children walked 25 yards from the zoo building to the cafeteria. For the first three months of the program there were no differences in behavior in the regular school classrooms between the children in the zoo program and the control group. Generalization to the schoolroom required between three and six months. At no time during the year study did the changes generalize to the children's residences. Even when decreases in symptomatology were observed in the school, symptom levels were always lower in the zoo program than in the school.

The limitation of the behavior change to the context defined by both the animals and the conventions of the companionable zoos may be illustrated by an event held at the Brandywine Campus during our initial study. Some of the children who had been working in the companionable zoo program were taken to the Philadelphia Zoo by their regular classroom instructor. At the zoo one of the boys was observed throwing stones at a group of flamingoes. The child was reprimanded, and the incident reported. When he was interviewed by an instructor back at the school campus and asked to explain his behavior, he said, "I wanted to feed them, but they didn't pay attention. Besides, what's it to you? They weren't our animals." The child was highly attracted to animals in both situations, but his apprehension of the rules for interaction was different in the two frames. This is also a good example of a lack of generalization of the moral stance learned in one situation to another.

Another example of the specificity of the zoo environment on moral behavior was the way that animals were incorporated into the threat and teasing behavior that was the normative pattern of interaction in the rest of the school. The students routinely threatened to kill or injure the animals of others with whom they were quarreling. They would pass notes containing such threats to one another in class. Yet no animal was ever injured in retaliation for some wrong or slight. In the zoo the same children were always solicitous about the welfare of all the animals. When an animal became sick or died, they were extremely supportive of each other. Yet all of this cooperative and sympathetic behavior stopped at the margin of the zoo program for the first six months of their exposure to the program.

Animals, Dialogue, and Sociability

The phenomena we observed in the zoo program are similar in most respects to the effects reported for transient interactions between people and pets in the society at large. Animals are said to act as social lubricants, to reduce social distance, and to facilitate social encounters. For example, people walking in public parks are more likely to be approached by strangers if they are in the company of an animal (Messent, 1983). When children who were confined to wheelchairs by handicapping conditions traveled with their dogs, they received 10 times as much social interaction as when they negotiated the same route alone (Mader, Hart & Bergin, 1989). In custodial institutions for the aged, visitation with animals results in increased attention to the environment, positive affect, and initiation of dialogue, even in patients who are socially withdrawn (Corsen & Corsen, 1981). Moreover, the staff are more interactive with their clients during and after volunteers and their dogs have visited the facility (Hendy, 1984). Residential animals have also been said to produce the same increase in social interaction and positive affect (Thomas, 1994). Children (Levinson, 1969) and adolescents (Peacock, 1986) in outpatient psychological treatment talk more freely and more volubly to their therapists. Autistic children display more social responses and less self-stimulation in the presence of animals (Redefer & Goodman, 1989). The reports of animals facilitating social interaction and dialogue almost uniformly describe the process as having two stages. First the patient directs her or his attention to the animal, and then the interaction generalizes to include the therapist. The animal can be said to be a guide or vehicle for bringing a socially isolated individual back into society.

The facilitation of dialogue by animals can be documented physiologically as well as behaviorally. Talking is almost invariably associated with elevation of blood pressure and heart rate, while listening, as long as the utterance is not threatening, results in lowered heart rate, blood pressure, and other signs of sympathetic arousal (Lynch, 1979). If blood pressure elevations are large, as when subjects are talking to auditors of perceived higher status, the surge of sympathetic activation can inhibit speaking. In general, those social conditions that reduce the stress response associated with talking facilitate dialogue. The blood pressure rises associated with talking to animals and talking in the presence of animals are lower than those associated with talking to people without the mediation of a pet animal (Katcher, 1981; Friedmann, Katcher, Thomas, Lynch & Messent, 1983; Baun, Bergstrom, Langston, & Thoma, 1984; Allen, Blascovich, Tomaka & Kelsey, 1991).

It can be inferred from the data on the influence of animals on social interaction that the animal alters social perception favorably. Why else would people with animals be more approachable than people without animals? However, there is also evidence that measured social attribution is positively influenced by the presence of animals in drawings (Lockwood, 1993) or pictures (Beck & Katcher, 1996) of people and animals. The use of animals in advertisements and the penchant of politicians for posing with pets are other indications that animals are a powerful means of positively altering social perceptions. In a variety of diverse situations the animal has the ability to irradiate people with trust, thus increasing their attractiveness along a variety of different but socially important dimensions.

It is important to note that the effects of interaction with animals cited above are transient and observed in the presence of the animal. There is no evidence that there is any relearning or reevaluation of responses toward people or any generalization of the effects to situations in which the animal is not present. The only exception is a report that children with pets are better at decoding human nonverbal emotional cues and are perceived more favorably by their peers (Guttmann, Predovic & Zemanek, 1985).

Animals as Guides into the Liminal State

How can we explain the favorable changes in self-image, social attractiveness, and social competence observed when animals are used in therapeutic situations with children? Or more generally, how do companion animals bring about similar changes in adults in therapeutic and social situations? First, it is necessary to make an obvious disclaimer. Snarling police dogs confronting demonstrators, charging horses bearing saber wielding Cossacks, or black bears ambling through New Jersey villages do not necessarily produce affable feelings toward animals or the people associated with them. The changes in human behavior we have described are seen only in certain, generally benign kinds of interactions. Within that set of interactions the changes in human behavior can be understood as a particular instance of a state that Victor Turner (1967, 1969, 1982) calls *liminality*. The hallmark of that condition is the intensification of good feelings and bonding between the participants that he calls *communitas*.

Turner (1982) developed the concept of liminality to describe a stage in rites of transition or passage. These have three phases—separation, liminality, and aggregation or return. Liminality is the state in which there is an intensification of lateral bonds (communitas)—acceptance of authority, minimization of differences between participants, commitment to task, increased sense of meaningfulness, and engagement in performance or play without role distance. It is during these periods of liminality that the behavior changes necessary for the passage into another state occur and the characteristics of the liminal state facilitate that behavior change. Later he applied the concept to situations that are not part of formal initiation rites—for example, the experience of being in a dangerous situation with others in combat, combating natural disasters, being caught in a blackout, or even participating in sport or attending theater. There are two aspects of liminality that are essential for understanding the interactions of children and animals—one social and affective and the other cognitive and normative. Turner (1982, p. 48) calls the social and affective aspect of the state *communitas*:

> Spontaneous communitas is a direct, immediate and total confrontation of human identities. A deep rather than intense style of personal interaction. It has something "magical" about it. Subjectively there is in it a feeling of endless power. Is there any of us who has not known this moment when compatible people—friends, congeners—obtain a flash of lucid mutual understanding on the existential level, when they feel that all problems, not just their problems, could be resolved, whether emotional or cognitive, if only the group which is felt (in the first person) as "essentially us" could sustain its intersubjective illumination.

> This illumination might succumb to the dry light of next day's disjunction, the application of singular and personal reason to the "glory" of communal understanding. But when the mood, style, or "fit" of spontaneous communitas is upon us, we place a high value on personal honesty, openness, and lack of pretentions or pretentiousness. We feel that it is important to relate directly to another person as he presents himself in the here-and-now, to understand him in a sympathetic (not an empathetic—which implies some withholding, some non-giving of the self) way free from the culturally defined encumbrances of his role, status, reputation, class, caste, sex or other structural niche. Individuals who interact with one another in the mode of spontaneous communitas become totally absorbed into a single synchronized fluid event.

He also relates that communitas does not erase structural norms from the consciousness of participants but symbolizes "the abrogation, negation, or inversion of the normative structure in which its participants are quotidianly involved" (ibid.).

The second general characteristic of liminality, its cognitive aspect, is described by Turner (1982) as a pedagogical system that proceeds by dissecting the symbols and relationships of a society into its parts and then recombining them in play satire and antinomian ritual (ibid., p. 26):

> Then the factors or elements of culture may be recombined in numerous, often grotesque ways, grotesque because they are arrayed in terms of possible or fantasied rather than experienced combinations—thus a monster disguise may combine human, animal, and vegetable features in an "unnatural" way, while the same features may be differently, but equally "unnaturally" combined in a painting or described in a tale. In other words, in liminality people "play" with the elements of the familiar and defamiliarize them. Novelty emerges from unprecendented combinations of familiar elements.

Elsewhere in the same volume he sees the essence of liminality in "the analysis of culture into factors and their free or 'ludic' recombination in any and every possible pattern, however weird" (ibid., p. 28). The activities within the liminal are described as "parody, abrogation of the normative system, exaggeration of rule into caricature or satirizing of rule" (ibid., p. 28).

To explain why animals can be agents of behavior and culture change, we have to recognize that people fantasize about animals, project human traits onto animals, and sometimes give superhuman powers to animals in a way that recombines, exaggerates, and contradicts human cultural elements. The animal can be compared to the masks that Turner describes as part of the ritual process: the combination of monstrous, human, and animal traits in masks, permits recombination of cultural symbols in new ways and is a source of illumination, play, creation, and recreation. It teaches how to dissemble the structure of social process and potentially rebuild it differently.

Under the right circumstances real animals can be agents for altering behavior the same way that animals in fairy tales alter the fate of the human children who encounter them in these stories. The child in the fairy tales meets in the forest an animal helper who has, at the very least, the human capacity for speech and who is encountered during a time when the child is between one condition and another. The forest with its dangers and helpful animals creates a special state akin to the state in rites of passage in which the participant is led out of one identity into another.

In some sense the educational institutions we observed could be looked on as organizations in need of a ritual process. Part of the problem with managing residences for highly aggressive children is that the children do not grant authority to the staff or to anyone else, for that matter. This contrasts with the absolute authority of the elders in the initiation rite. The children, unlike cadets at West Point, do not tend to affiliate with each other but remain fiercely antagonistic to their peers. There is no evident communitas or feeling of coming through a difficult situation together. There is also no ritual content. The children are not expected to master any doctrine or any explanation of their state; they are expected only to learn how to conform to the behavior control system. The presence of the zoo program introduces a liminal state within that limited environment. The animals and the children's play with them create a state of liminality, and the moral climate set by the instructors gives a shape to their experience. The state is, by and large, temporary and context specific. During the period of observation the behavior did not quickly extend to the rest of the institution. Only after three to six months of interaction in the liminal zoo environment was there evidence that something had been learned that could be imported outside of the zoo into the regular school rooms.

The ability of children to learn new patterns of social behavior by first reasoning about real animals and then applying those social insights to people suggests that there may be more than one reason that animals appear in morality tales for children. The conventional explanation, following Lévi-Straus (1968), is that the animal acts as a visible embodiment of behavioral or temperamental characteristics that are not so visible in human beings. Cowardly and brave human beings may look alike in the supermarket, but lions and jackals always look different. Our experience suggests that animal are the bearers of social instruction because those precepts are then insulated from the harsh realities of real life. Most of the children we treated came from homes in which they were neglected and abused. They learned to expect anger and aggression and projected those feelings onto almost all of their social contexts. They agreed with Mack the Knife that "The world is mean and man uncouth." Although these children read, heard, or viewed the same fairy stories, the same fables, and the same television programs with cute talking animals that we all did, they knew that the people around them did not play by those rules, and in retaliation they didn't either.

When moral precepts are generated by stories about animals, they can never be contradicted by experience within the family. The moral or social code is insulated from daily life. The models for friendship, devotion, support, love, and sacrifice cannot be contradicted by the child's disappointing, demeaning, or brutal experience within the family (Cartmill, 1983). The presence of animals—especially animals that the child is not familiar with—suggests a suspension of rules learned from daily life. That suspension creates the field in which the liminality of the animal experience develops.

The ability of animals to serve as exemplars for particular virtues (or vices) in myth and fairy tale also suggests how animals help people enter a liminal condition and thereby how the feeling of communitas is constructed. To understand human-animal relationships it is necessary to introduce the concept of *split objects*, which are a derivative of the infant's division of people into familiar and comforting on one hand and strange and dangerous on the other (Ogden, 1990). There is, in some more than others, a lifelong tendency to use internal representations or templates to make rapid decisions about others, separating them into the "all good" and the "all bad." Racial and ethnic stereotypes, the process of falling in and out of love, and much political and religious imagery are all based on the formation of split images. What I would like to suggest is that owners of companion animals transform their animal into a pet by using it as a vehicle for projection of a good split image of the self as child or mother (or both). The pet is an amalgam of a real animal and a split image. That is what is meant when the pet or animal is said to be a transitional object. Winnicott (1971), who more than any other psychoanalytic theorist knew the value of play in human life, talked of a transitional relationship that both created and transcended the play with the stuffed toys and security blankets of early childhood. In a transitional relationship a child or an adult takes the attributes of a purely subjective object—a fantasy object—and projects them onto some real entity in the external world. In the transitional relationship the play is the movement of attributes of fantasy objects from an internal space to real objects in external space.

The idea of the animal as a transitional or split object can be illustrated with Perin's (1981) description of modern urban Americans' conceptualization of their relationships with pet dogs. Perin's (1981, p. 77) characterization of the superabundant love of the dog—a kind of feeling that the children in our programs expressed toward a variety of animals and that had powerful reflections onto their behavior toward each other and their instructors—closely resembles the interpersonal correlates of the liminal state that Turner calls communitas:

> The quality that best characterizes the bond of feeling between people and dogs is abundance; in fact, superabundance. For the bond is often seen to represent an *excess of love* having no rightful place in human relationships, supersaturated feelings people are not able to or not allowed to bestow on other people.

She goes on to note that these feelings translate into idealization and casts about for a template elsewhere in human experience (ibid., 1981, p. 81):

> When else have we ever actually received unquestioning devotion, utter adoration, a total absence of judging, unspeakably overwhelming trust, unspoken understanding and unbounded love? How does it come to be that for our "best friend" we turn to another species? That the fulfillment of such supersaturated expectations may be anticipated only from another species brought the realization that I was in the presence of something transhuman or metaphysical—that is, a symbol, a condensation of meanings richer than real.

Animals make good transitional beings because they move and show intentional behavior, behaving more like a person than a stuffed toy.

Unlike stuffed toys, which provide only passive soft touch, animals are capable of giving active affection and seeking out the child. But most important, they never can contradict the attributes projected on to them with words. Nor can the animal alter the relationship by redefining the child's position with words. Throughout our entire lives, our animals are there as transitional objects, being what we imagine them to be, serving as vehicles for projecting those admirable traits that we find so lacking in fellow human beings. They even serve with their coat of shining virtues to redefine by contrast the uncertain and amoral world of human companions.

With the idea of split objects we also have an explanation for the good feelings that permeate those liminal moments when we feel so close to our fellow man. This is the formation of split objects. We construct those very good fellows by the process of addition of the good and elision of the bad. Like our pets, those people we bond with during those moments of communitas are perhaps too good to be true or too good to last beyond the moments when we are overwhelmed with the feelings generated by their palpable presence.

Conclusions

When therapeutic education was structured around contact with animals, we observed a decrease in aggression and negative social attribution and an increase in cooperative behavior and social affiliation. Although there was some generalization of these improved behaviors to the children's regular classrooms within six months of treatment, these positive changes were, by and large, limited to the times when the children were experiencing guided contact with animals and nature. The children's manner of behaving when in contact with animals strongly resembles the state Victor Turner describes as liminality and communitas. The concept of liminality permits the conceptualization of the therapeutic effect of animals as a particular instance of a more general phenomenon seen in many different kinds of situations that do not contain animals—rites of passage, emergencies, and risky situations (such as those provided by sports).

Suggesting that animals can induce a liminal state does not explain why they can. Possibly, to some extent, animals are always conceptualized as transitional objects—beings apprehended through both reality testing and the projection of fantasy. When those attributed (rather than observed) characteristics were positive, as they almost always were, in the situations we observed, then the benign social attractiveness characteristic of communitas was generated and then generalized from the animals to peers and teachers.

Moreover, animals were readily perceived as positive transitional objects because most children who found their way into treatment had been injured by human beings, had control issues with people, and had been assaulted and controlled with words. The absence of speech and obvious controlling strategies in animals permitted them to serve as appropriate vehicles for the projection of positive feelings. These positive feelings, as noted, then generalized to people. Under the proper circumstances and guidance, the novel and playful ways that social attributes are recombined in the liminal state (generated between people and animals) destabilize ingrained patterns of thinking and permit the learning of new behavior patterns.

These results have strong implications for the understanding and shaping the values that children place on animals and nature. The study of children's values could profit from designing experiments in which interviews are conducted with and without the presence of animals or natural settings. If, as we have suggested, the child's level of arousal, self-concept, and social competence are all effected by the presence of animals, then testing him or her without animals may decrease the experimenter's ability to predict how the child's values would be actualized in behavior toward animals. Alternatively, the presence of an animal would be expected to alter the nature of the relationship between experimenter and subject and thus change the demand characteristics of the experiment. Indeed, testing children in both conditions—with and without the presence of an animal—would be an elegant means of testing for experimenter biasing effects. Those effects would be expected to be stronger in the presence of animals or nature because under those circumstances the subject would be more highly motivated to please the experimenter.

Liminal states induced through contact with animals also have implications for the problem of humane education. The data suggest that conducting human education in the presence of animals or natural settings is more likely to result in acceptance of the values of the educators and facilitate behavior change.

References

Allen, K. M., Blascovich, J., Tomaka, J., & Kelsey, R. M. (1991). Presence of human friends and pet dogs as moderators of autonomic stress in women. *Journal of Personality and Social Psychology, 61*, 582–589.

Baun, M. M., Bergstrom, N., Langston, N. F., & Thoma, L. (1984). Physiological effects of human/companion animal bonding. *Nursing Research, 33*, 126–129.

Beck, A., & Katcher, A. (1996). *Between pets and people: The importance of animal companionship*. West Lafayette, IN: Purdue University Press.

Cartmill, M. (1983). "Four legs good, two legs bad": Man's place (if any) in nature. *Natural History, 11*, 65–78.

Corsen, S. A., & Corsen, E. O. L. (1981). Pet animals as bonding catalysts in geriatric institutions. In B. Fogle (Ed.), *Interrelations between pets and people* (pp. 146–174). Springfield, OH: Thomas.

Fine, A. (2000). Animals and therapists: Incorporating animals in outpatient therapy. In A. Fine (Ed.), *The handbook on animal assisted therapy: Theoretical foundations and guidelines for practice* (pp. 179–212). New York: Academic Press.

Friedmann, E. (2000). Animal-human bond: Health and wellness. In A. Fine (Ed.), *The handbook on animal-assisted therapy: Theoretical foundations and guidelines for practice* (pp. 41–55). New York: Academic Press.

Friedmann, E., Katcher, A. H., Lynch, J. J., Thomas, S., & Messent, P. R. (1983). Social interaction and blood pressure: Influence of animal companions. *Journal of Nervous and Mental Disease, 171*, 461–465.

Guttmann, G., Predovic, M., & Zemanek, M. (1985). The influence of pet ownership on non-verbal communications competence in children. In *The human-pet relationship* (pp. 58–63). Vienna: Institute for Interdisciplinary Research on the Human-Pet Relationship.

Hendy, H. H. (1984). Effects of pets on the sociability and health activities of nursing home residents. In R. K. Anderson, B. L. Hart & L. A. Hart (Eds.), *The pet connection* (pp. 430–437). Minneapolis: University of Minnesota.

Katcher, A. (1981). Interactions between people and their pets: Form and function. In B. Fogle (Ed.), *Interrelations between people and pets* (pp. 41–67). Springfield, OH: Thomas.

Katcher, A. (2000). Animal-assisted therapy and the study of human-animal relationships: Context or transitional object? In A. Fine (Ed.), *The handbook on animal-assisted therapy: Theoretical foundations and guidelines for practice* (pp. 461–474). New York: Academic Press.

Katcher, A., & Wilkins, G. (1993). Dialogue with animals: It's nature and culture. In S. Kellert & E. O. Wilson (Eds.), *The biophilia hypothesis* (pp. 173–200). Washington, DC: Island Press.

Katcher, A., & Wilkins, G. (1998). Animal-assisted therapy in the treatment of disruptive behavior disorders. In A. Lundberg (Ed.), *The environment and mental health* (pp. 193–204). Mahwah, NJ: Erlbaum.

Katcher, A., & Wilkins, G. (2000). The centaur's lessons: Therapeutic education through care of animals and nature study. In A. Fine (Ed.), *The handbook on animal assisted therapy: Theoretical foundations and guidelines for practice* (pp. 153–178). New York: Academic Press.

Katcher, A., & Wilkins, G. (2001). *The Centaur's lessons: The companionable zoo method of therapeutic education based upon contact with animals and nature study*. Chicago: PAN-ATA Press.

Kaye, D. M. (1984). Animal affection and student behavior. In R. K. Anderson, B. L. Hart & L. A. Hart (Eds.), *The pet connection* (pp. 101–104). Minneapolis: University of Minnesota.

Levinson, B. (1969). *Pet-oriented child psychotherapy*. Springfield, OH: Thomas.

Lévi-Strauss, C. (1968). *The savage mind*. Chicago: University of Chicago Press.

Lockwood, R. (1983). The influence of animals on social perception. In A. Katcher & A. Beck (Eds.), *New perspectives on our lives with companion animals* (pp. 64–71). Philadelphia: University of Pennsylvania Press.

Lynch, J. (1979). *The broken heart: The medical consequences of loneliness*. New York: Basic Books.

Mader, B., Hart, L. A., & Bergin, B. (1989). Social acknowledgments for children with disabilities: Effects of service dogs. *Child Development, 60*, 1528–1534.

Messent, P. (1983). Social facilitation of contact with other people by pet dogs. In A. Katcher & A. Beck (Eds.), *New perspectives on our lives with companion animals* (pp. 37–46). Philadelphia: University of Pennsylvania Press.

Ogden, T. (1990). *The matrix of the mind*. Northvale, NJ: Aronson.

Peacock, C. (1986). The role of the therapeutic pet in initial psychiatric sessions with adolescents. Paper presented to the Delta Society International Conference, Boston, August.

Perin, C. (1981). Dogs as symbols in human development. In B. Fogle (Ed.), *Interrelations between pets and people* (pp. 68–88). Springfield, OH: Thomas.

Redefer, L. A., & Goodman, J. F. (1989). Brief report: Pet-facilitated therapy with autistic children. *Journal of Autism and Developmental Disorders, 19*(3), 461–467.

Ross, S., Vigdor, M. G., Kohnstamm, M., DiPaoli, M., Manley, B., & Ross, M. (1984). The effects of farm programming with emotionally disturbed and handicapped children. In R. K. Anderson, B. L. Hart & L. A. Hart (Eds.), *The pet connection* (pp. 120–130). Minneapolis: University of Minnesota.

Tharp, R. G., & Gallimore, R. (1990). *Rousing minds to life*. Cambridge: Cambridge University Press.

Thomas, W. (1994). *The eden alternative: Nature, hope and nursing homes*. Sherburne, NY: Eden Alternative Foundation.

Turner, V. (1967). *The forest of symbols: Aspects of Ndembu ritual*. Ithica, NY: Cornell University Press.

Turner, V. (1969). *The ritual process*. Chicago: Aldine.

Turner, V. (1982). *From ritual to theater*. New York: Performing Arts Journal Publications.

Vygotsky, L. (1986). *Thought and language*. Cambridge, MA: The MIT Press.

Winnicott, D. W. (1971). *Playing and reality*. London: Tavistock.

8

Spots of Time: Manifold Ways of Being in Nature in Childhood

Louise Chawla

Place and Time

I am beginning this chapter in the Lake District of England, in an eighteenth-century house with a Victorian addition, not unlike the house William Wordsworth lived in as poet laureate of Britain. Outside my window are towering pine trees whose tallest crowns reach to the zenith of the sky. Beneath them are the massed greens of rhododendrons, the tangled red-hipped branches of roses, the rotted clumps of last year's flower beds, and the spongy silver-green grass that leads down to a meadow where sheep graze, which leads to a marsh, which leads to an inlet of Lake Windermere, which leads to the mountains that rim Hawkshead, site of Wordsworth's schooling as a boy. By one of life's synchronicities, just as I was due to begin this chapter, I was invited by friends to this region where Wordsworth fashioned the topic of children and nature into a significant modern theme.

Last night, the close mist that had shrouded all of the previous day suddenly lifted, leaving a night so clear that the blue-black of the sky or the dust of the Milky Way could be alternately figure or ground, so bright and myriad were the stars. Standing here, in this place where the Lake District authors were the first to rally against the Industrial Revolution's accelerating destruction of the natural universe, this point in place and time appeared a fair metaphor for the bridge that those of us now living need to make between our present efforts to heal the planet's wounds and the visions of possible harmonies between humanity and nature, and our past and present selves, that the Romantics delivered to us.

In the Prelude, Wordsworth (1850/1971, 12.208–12.218) called such moments of clarity "spots of time":

There are in our existence spots of time,
That with distinct pre-eminence retain
A renovating virtue . . .

They are moments that merit our return and meditation. Many especially resonant spots of time, Wordsworth observed, date from childhood.

This chapter reviews Romantic ideas that first defined the modern theme of childhood and nature and contributed to the hermeneutic tradition of research to which this chapter belongs. It introduces the ideas of the Swiss philosopher Jean Gebser, who extended this tradition, as a framework for describing different ways of knowing nature in childhood, as well as different ways of relating to childhood in adulthood, with an emphasis on dimensions of experience that have received limited research attention. "Nature," for the purposes of this chapter, refers to the "green world" of forests, fields, farms, parks, and gardens—the elements of earth, water, air, and growing things that exist independent of human creation, although they may be shaped into forms of human design. Children's relations with animals figure here, but I leave this aspect of the topic to other authors in this book. The chapter closes by reflecting on where research has currently brought us in our efforts to understand how children develop different ways of knowing and being with respect to the natural world.

Romantic Legacies

From here within the Lake District, the towering oaks and pines, the mountains, the meadows, the early winter darks, and the long summer light are immediate presences. This region has been transformed by human activity for millenia, yet these elements of nature dominate the human scale. To try to communicate their qualities, one reaches for poetry, which has a music that approximates the harmonies of this landscape. In trying to express how growing up here as a boy influenced him, Wordsworth (ibid., 1.464–1.475) said it in this way:

Ye Presences of Nature in the sky
And on the earth! Ye Visions of the hills!

And Souls of lonely places! can I think
A vulgar hope was yours when ye employed
Such ministry, when ye through many a year
Haunting me thus among my boyish sports,
On caves and trees, upon the woods and hills,
Impressed upon all forms the characters
Of danger or desire; and thus did make
The surface of the universal earth
With triumph and delight, with hope and fear
Work like a sea?

Wordsworth thought of nature as something ensouled, possessing real characters "of danger or desire." Unlike Romantic philosophers like Fichte or Hegel, who argued that the world is entirely a creation of the mind, Wordsworth usually wrote of nature as something with an independent reality of its own that impresses its characters on our senses, to which we respond with more or less depth of feeling and reflectiveness.

So it was for him with memory as well. Memory, for Wordsworth, was another presence that impresses its characters on us and that we half create and half receive. It is a second remove from nature: the characters of nature mark our memories, which have a reality of their own that we carry with us, forming resources or risks that we later draw on.

Wordsworth attributed many legacies to childhood experiences of nature. In the beginning, he conjectured, a baby transfers the love that it receives from its mother to surrounding things. When its tendencies to love, pity, and respond to another are nurtured by responsive caretaking, a small child is prepared to interact with the objects of the natural world with creativity and sympathy. Playing on the ambiguity of the Neoplatonic tradition—in which the body is the prison of the soul and yet the material world reveals patterns of divinity—Wordsworth (ibid., 2.258–2.260) called such a receptive child "an inmate of this active universe," whose senses

Create, creator and receiver both,
Working but in alliance with the works
Which it beholds.—

In this way, Wordsworth believed, early childhood prepares a foundation for a responsive give and take with the outer world throughout life. (For discussions of comparable ideas in contemporary psychoanalytic theory, see Holmes, 1999, and Searles, 1959.)

Many childhood impressions, Wordsworth thought, remain obscure and inarticulate, leaving the soul "remembering how she felt, but what she felt / remembering not" (ibid., 2.315–2.317). Nevertheless, Wordsworth (1798/1952) believed that when people have learned to respond to the world with sympathy in childhood, then childhood impressions of nature contribute to "tranquil restoration" in later life and have a moral influence that encourages "little, nameless, unremembered acts / of kindness and of love" ("Lines Composed a Few Miles Above Tintern Abbey," lines 30, 34–35). They also create a habit of absorption in the natural world, when, "with an eye made quiet by the power / of harmony, and the deep power of joy / we see into the life of things" (ibid., lines 47–49). If people are fortunate, one of the most important possibilities of feeling that childhood passes to adulthood is joy.

Another habit of childhood is to animate nature and invest it with moral significance. In *The Prelude* (3.132–3.135), Wordsworth recalled his own perceptions of all natural forms—"rock, fruit or flower":

. . . I saw them feel,
Or linked them to some feeling: the great mass
Lay bedded in a quickening soul, and all
That I beheld respired with inward meaning.

Because he thought of the universe as a living whole, to him this sense of a moral life in nature was a true insight.

In contemporary social science, Romantic connections between childhood and nature are usually dismissed as "romantic" in the most pejorative sense—an idealized, unrealistic picture of unbroken innocence and happiness that represses the reality of social discord and pain. What this cursory dismissal actually expresses is ignorance of Romantic thought. For Wordsworth, as for other major Romantic writers, childhood was a "fair seed-time" for his soul in which he grew up "fostered alike by beauty and by fear" (ibid., 1.301–1.302). For example, to illustrate his own childhood "spots of time," Wordsworth described when he was not yet six and discovered the site on the moor where a murderer had been hung, and later at the age of 13, sitting on a misty crag, waiting impatiently for the holidays from boarding school: a holiday time, as it turned out, when his father died (ibid., 12.226–12.335). Wordsworth believed that meditating on memories of fear, as well as delight, may be renovating.

Dismissing Wordsworth on these partial grounds evades a serious confrontation with one of the main elements of his argument. According to his friend Samuel Taylor Coleridge, Wordsworth's mission as a poet was to offer an alternative to the mechanistic view of nature that characterized empiricist philosophers like Locke and Hobbes and the bold new world of secular science, for whom nature was a machine devoid of spirit or moral significance (Coleridge, cited in Abrams 1971, p. 145). According to these "Mechanic Dogmatists," in Coleridge's term, our proper relation to nature is to decipher its laws by registering its activity as passively and accurately as possible through our senses and analyzing these facts through our reason, enabling us to control and manipulate nature for our comfort. Useful as this instrumental relationship may be, Wordsworth (1798/1952) maintained that we more wisely conceive nature to be a living organism to which we rightfully feel bound by love and fear ("Lines Composed a Few Miles Above Tintern Abbey").

This argument is not merely academic. Wordsworth drew into his reflection different strands of Western philosophy and religion that continue to pervade discourse about children and nature, whether people use these ideas with a conscious awareness of their sources or not: the Gospel idea of the redeeming child who represents the state of paradise, the Augustinian child of "original sin" who is born to knowledge of pain and death, the Platonic vision of the universe as a beautiful ensouled creature, and St. Bonaventure's metaphor that the universe is a book in which the initiated can read divine mysteries and moral teachings (the "characters" of nature in a double sense) (Chawla, 1994b). Wordsworth translated these traditions into secular form to salvage a belief in the moral significance of his memories and nature itself.

In my experience, when I have talked with men and women about how they use their childhood memories of nature, I have found that the dilemma that Wordsworth faced remains a contemporary one: those who accept the standard of scientific rationalism that nature is an amoral mechanism consider their early memories of an animated world childish nonsense or simply irrelevant. Only those who believe nature to be a living whole with intrinsic meaning find childhood a period of insight (Chawla, 1994b).

A Hermeneutic Heritage

Despite the tendency of twentieth-century social science to dismiss Romanticism as a period of superficial idealism, across the span of the twentieth century the social sciences were increasingly drawn into the dilemmas of a quintessentially Romantic question: How do I define my own self-consciousness, and from this perspective, how do I know an Other? This question rings throughout contemporary debates about the philosophy of science (Guba, 1990).

In the terms of Ricoeur (1981), this dilemma about the meaning of the self and the world is manifested in a division between a "hermeneutics of suspicion" and a "hermeneutics of recollection." The first seeks to expose how the meanings of things are determined by underlying drives for personal power or social and economic control. To the degree that it is appropriate to speak of reality, it lies in these dynamics. As Kahn (1999) has noted, when this theory is applied in the form of postmodern deconstruction, it invites nihilism and opportunism in relations with people, animals, and the environment. Certainly, by this view, Romantic ideas about childhood and nature have to be dismissed.

A hermeneutics of recollection, by contrast, seeks to uncover a plenitude of coexisting meanings. It is here, in this tradition, that this chapter is positioned. Here, Wordsworth belongs to the history of this chapter's method as well as to its topic, for this form of hermeneutics descends from a conserving line of Romantic thought.

Modern hermeneutics begins with Friedrich Schleiermacher, a German theologian who was closely associated with Friedrich Schlegel and other German Romantics. In 1798 Wordsworth traveled through Germany to familiarize himself with German Romanticism. He wrote the first extended version of *The Prelude* and the first and second editions of the *Lyrical Ballads* during the same years that Schleiermacher drafted the main principles of his work. Both men posed anew the central question of hermeneutics: How do I know and interpret the voice of another, whether this other is a text of classical philosophy, a passage of scripture, or a body of law (the subjects of Schleiermacher's study) or an object of nature, a period of memory, a child, or any other human being (themes of Wordsworth's writing)? Both men concluded that one method that we can use is to imaginatively project ourselves into another and sympathetically know another, in effect, from the inside (Bruns, 1992). In Wordsworth's words, we can do so because there is a real "filial bond" that connects us to other people and things in the universe, past as well as present—in Schleiermacher's astounding phrase, because "each person contains a minimum of everyone else" (Bruns, 1992, p. 161).

By the early twenty-first century, this trust in our human ability to identify with another has given way to a widespread disbelief that any universal principles connect human beings or bind them to the universe or even less that the universe is ensouled. Instead, it is common discourse now that ideas about nature, childhood, or anything else are social and cultural constructions imposed on an otherwise meaningless world (see discussions in James, Jenks, & Prout, 1998; Soule & Lease, 1995). Suspicions prevail.

This chapter is directly founded on the work of Hans-Georg Gadamer, who labored to secure a foundation for the social sciences in the midst of this controversy. Gadamer (1975), like Wordsworth, proposed that we half create and half receive the world. He drew on the ideas of his teacher, Heidegger (1949), who drew in turn on the phenomenologist Husserl (1977), who argued that we need to self-consciously "bracket" the inescapable presuppositions and biases of our horizon in time and place so that we can perceive how things present themselves. In Heidegger's terms, despite our embeddedness in our history, we can create a "clearing," metaphorically, in which we invite a phenomenon to show itself on its own terms—whether child, tree, or any other object of knowledge. In this shared space, we seek to move from a mode of domination to a mode of listening.

What happens in this case, according to Gadamer (1975), is a "fusion of horizons" in which we are transformed by the encounter: we understand ourselves and what matters to us further in the light of the object we encounter, even as we understand this object in the light of ourselves and our interests. According to Gadamer, we need to understand the pursuit of the human sciences as participation in this ongoing process of encounter. This chapter is written in this tradition, with a belief that our subjects of study—such as a child in nature—have realities of their own and therefore that our role as researchers is to listen and observe as openly as possible, even as we must be sensitive to how our own qualities and presuppositions shape our research.

Together, the hermeneutic tradition and the Romantic interest in childhood pose essential questions to anyone who takes up the theme of children and nature: What is our connection as adults to children? To nature? What is the significance of this subject for us? How do we know it according to our particular place and time in history? How is the study of this subject embedded within our practical interests?

For myself, as I began research into childhood memories of nature as well as children in the environment, I found that the topic I had chosen posed special quandaries. People's memories of their childhood, as well as children's discourse about nature, sometimes suggest realms of experience that the dominant practice of science considers inadmissible on its own terms. Rather than diminishing what I heard by labeling it primitive, childish, or "romantic," as the social sciences have tended to do, I looked for a system of thought that could contain it—and found it in the work of Jean Gebser, a Swiss philosopher who worked in the phenomenological and hermeneutic traditions and whose thought has special relevance with regard to our relationship to nature, childhood, and different moments of our lives. Gebser has provided a vocabulary to talk about otherwise difficult-to-acknowledge aspects of children's experience of the natural world. Therefore the following sections briefly outline some of his major ideas.

The Ever-Present Origin

As someone who crossed the boundaries of nations and disciplines in his life, it is not surprising that Jean Gebser crossed boundaries between different ways of being in the world in his thought. Born in the Polish region of Prussia in 1905, Gebser lived the peripatetic life of many European intellectuals during the rise of fascism and World War II. He successively fled fascism from Germany to Italy, to Spain, to France, to a haven at last in Switzerland, which became his base for postwar lecture tours in Europe, the Americas, and Asia. Even a very partial list of the friends with whom he worked during these years suggests the breadth of his interests and experience—the poet Federico García Lorca, the psychologist Carl Gustav Jung, the biologist Adolf Portmann, the physicist Werner Heisenberg, the spiritual leader Lama Anagarika Govinda. As an approximate title for his interests, the University of Salzburg in Austria created a chair in comparative cultures for him. He died in 1973 (Feuerstein, 1987, 1989).

Nature, for Gebser, may be best translated by the Greek *physis*: an upwelling of self-organizing energy that pours itself forth in all the forms of the universe—an ever-present origin, physically and spiritually. Therefore, Gebser's sense of nature transcends the "green world" that is the topic of this chapter, but his ideas about human relationships with nature in this all-encompassing sense are relevant to this chapter's focus on nature in a more narrow sense. According to Gebser's most ambitious text, *The Ever-Present Origin*, different cultures, periods of human history, and moments in an individual's life exhibit different dominant *Gestalten*, or structures of consciousness through which the world is organized and experienced. Gebser accepted Kant's argument that we impose a structure of time and space on the world, but he proposed that our human nature carries the potential for five definably different structures (Mickunas, 1973).

I came to Gebser's ideas because he provides a way out of the hierarchy of "primitive" and "prelogical" thought versus "advanced" logical thinking that pervades developmental psychology and that has particularly influenced discussions of children and nature (Chawla, 1994b). In the descriptions of different space-times that follow, readers may hear echoes of the idea that "ontogeny recapitulates phylogeny"—the popular nineteenth-century idea that children play out different phases of human evolution (Gould, 1977), but Gebser rejected the linearity of this notion through three radical departures (Chawla, 1993):

- Although he noted that certain ways of structuring time and space may dominate during certain periods of human history or ages of an individual, each structure remains an ever-present potential of our human nature.

- No space-time structure is inherently inferior or superior to another, but each is to be evaluated according to its own effective and defective possibilities.
- In addition to familiar *mental* and *mythic* forms of consciousness, Gebser defined preverbal *archaic* and *magic* forms, to which he gave equal significance. A fifth form, which he termed *integral*, maintains openness to the effective possibilities of the other four. Because of its flexible movement through different structures of space and time, integral consciousness serves as a goal for development.

This new conceptual context encourages us to value what has been previously termed "lower" childhood ways of being in the world as much as "higher" adult ways, to evaluate the consequences of each for individuals and the life around them, and to maintain continuity with each potential.

In the sections below, brief summaries of Gebser's descriptions of the four basic forms of consciousness are followed by discussions of their relevance to existing research on children and nature. In conclusion, the chapter reviews Gebser's concept of integral consciousness as a way of preserving what is most advantageous in all of these different ways of being, if we are to resolve the intensifying environmental challenges that we face.

The Necessity of the Archaic

In his description of archaic consciousness, Gebser (1985, pp. 43–45) noted that *arche* is Greek for "origin" and he refers to the archaic structure as "identical with origin" and therefore the "wisdom of origin,"—when people do not yet differentiate themselves from their surroundings. It is the consciousness of animals and probably the dominant consciousness of infants, which we may reenter in later life in repose and reverie, when we are simply absorbed in our body and our place. It includes the operations of our autonomic nervous system, which are usually not thought of as a form of consciousness and yet are a vital level of awareness and exchange with the environment, even though it cannot be articulated in words.

Rather than considering this foundation of consciousness "merely physical," Gebser observed that it is our level of real identity with the world around us and therefore that it forms our primal connection to nature and to human wisdom regarding our part in nature. Even at this level, however, it is a foundation for trust or fear. "Life is forever menaced by chaos," Gebser (1962, p. 6) observed, "and must restore balance with every intake of breath."

Archaic experience is elemental in an immediate physical sense. It is a way of being baptized in the world by immersion, such as children in play who literally live close to the ground and up against the full sensory qualities of things—making hiding places under tables and bushes, climbing trees, rolling down hills, squatting in mud and water, and peering under rocks, surrounded by smells, textures, and details that adult height and habits will later remove them from. It is possible to become absorbed in elements of the built environment in this way, but this is absorption in products that do not give life; whereas Gebser's reference to archaic consciousness as "identical with origin" suggests experiences of identity with elements like rock, earth, water, light, and leaf out of which human life in fact comes. The concept of the archaic implies, in turn, that a child's receptivity to these elements on terms of trust depends on the condition of its body. A child nourished by love, food, and rest knows the world differently than one anguished by hunger and insecurity.

Magic Union

Archaic consciousness eludes expression in words. Therefore, it is rarely acknowledged as a form of consciousness at all in the word-centered world of psychology. Magic consciousness defies rational explanation. Therefore, it also tends to be ignored or dismissed as "irrational" by the rational world of research. For Gebser (1985, pp. 45–60), however, magic consciousness is a vital experience that apprehends the power of our connection with the world. In contrast to archaic "identity," Gebser wrote about magic "union" with the world, as "union" implies a self-aware coming together of self and other. Magic is commonly associated with chant and ritual, but as Gebser described it, it is a silent intuition of the world's power and our own power. Therefore, it is an unforgettable experience that, in Gebser's words (1985, p. 251), "simultaneously 'realizes' the unity of the world and the fundamental unity of the individual with the world."

These different elements of Gebser's description may be illustrated by the following memory by a British woman, as she recalled an episode that she could date to the age of five, when she and her mother were walking over the moors as the sun declined, the chill of evening came on, and a mist formed over the ground (quoted in Robinson, 1983, p. 33):

> Suddenly I seemed to see the mist as a shimmering gossamer tissue and the harebells, appearing here and there, seemed to shine with a brilliant fire. Somehow I understood that this was the living tissue of life itself, in which that which we call consciousness was embedded, appearing here and there as a shining focus of energy in the more diffused whole. In that moment I knew that I had my own special place, as had all other things, animate and so-called inanimate, and that we were all part of this universal tissue which was both fragile yet immensely strong, and utterly good and beneficent.

In Gebser's terms, this account evokes an effective experience of magic. The woman attributed to this memory "a kind of reservoir of strength fed from an unseen source, from which quite suddenly in the midst of the very darkest times a bubble of pure joy rises through it all, and I know that whatever the anguish there is some deep centre in my life which cannot be touched by it" (ibid., p. 33).

Gebser (1985, p. 51) noted, however, that magic awareness of our own and the world's power also opens us to the possibility of fear: "fear that man is compelled to rule the outside world—so as not to be ruled by it." In this defective form, he found magic consciousness prevalent in the modern world's obsession with machines and technologies and in its limitless drive to transform all of the earth's resources into the objects of human desire.

Mythic and Mental Experience

With the mythic and the mental, we come to two familiar forms of consciousness that are accessible through language and symbol. The mythic structure, as Gebser (1985, pp. 61–73) described it, gives voice to the powers of empathy, sympathy, and associative thinking that magic consciousness makes possible and communicates a collective sense of "ours" and "us." Gebser (1985, p. 65) noted that the words *myth* and *mouth* come from the Sanskrit root *mu*, "to sound." As the sound of the mouth reveals emotions, it gives consciousness access to an internal life of sensibility and imagination.

Gebser observed that whereas magic experience involves a sense that all time and space are concentrated *here*, myth takes place in cyclic time—to the rhythm of the in-breath and out-breath, the heartbeat, day and night, the seasons, and the generations, as these rhythms are expressed in the circling patterns of the dance, song, poetry, ritual, and the visual arts. Rather than dualities, mythic consciousness notices "complementarities," or distinctions within relationships, such as child and parent, man and woman, light and dark, earth and sky, spring and autumn. Applied to places of strong personal feeling and group identity, it expresses a sense of sacred place. In defective forms, it becomes propaganda and empty ritual.

In contrast to myth, mental time and space are structured by an observing I/eye that assesses the environment objectively, evaluates it in the abstract, and measures it rationally and often quantitatively. In Gebser's description (1985, pp. 73–97), the ruling term of this form of consciousness is *ego*, "I," and the ruling sense is sight. It makes possible perspective, paradox, abstraction, rational reflection, and self-assertion—all of which may take effective or defective forms. At its best, it includes a healthy sense of self-efficacy and self-esteem and illuminative powers of focus and insight, but it creates precarious dualities—self versus other, subject versus object, man versus nature, adult versus child. Space is perceived in three dimensions oriented to the one-point perspective of the observer's lines of sight, and time becomes an "arrow"—an irreversible quantified line. At its worst, mental-rational powers of calculation combine with a defective egoism to reduce nature and other people to mere mechanisms to manipulate and consume.

Belonging to the mental structure as the sciences do, it is not surprising that research regarding children and nature has overwhelmingly emphasized cognition through this form of consciousness, with a focus on environmental reasoning, knowledge, and attitudes (Kahn, 1999; Wals, 1994; Zimmerman, 1996). Important as knowledge, reasoning, and attitudes are as children seek to understand the world and their place within it, they do not form our deepest levels of connection with the natural world. The qualities of our attention and movement through the sensory world, our sense of agency and identity, and the play of emotions they engender are at least equally important. As Hungerford and Volk (1990, p. 11) note in their review of predictors of responsible environmental behavior, in addition to knowledge, attitudes, and skills necessary to take action, people are moved to protect the environment through "environmental sensitivity," a term that they define as "an empathetic perspective toward the environment" and associate with positive experiences in natural areas. Therefore, the remainder of this chapter turns to research that can be related to Gebser's archaic, magic, and mythic structures of consciousness, where identification and attachment to the natural world form a basis for this "empathetic perspective."

A Foundation for Archaic Identity

Gebser's description of archaic consciousness suggests the importance of time spent in the natural world in an unthreatened way that encourages a bond of connection rather than fear. It illuminates a recurring finding of studies that ask people to identify the sources of their interest in the natural world or commitment to protect it. Most frequently, people mention natural areas that they frequented in childhood or adolescence through play, hiking, camping, fishing, or other routine outdoor activities (Chawla, 1998; Palmer & Suggate, 1998). In some studies, family members are also frequently mentioned.

In my own work, when I interviewed 56 environmentalists in Kentucky and Norway about their motives for protecting the environment, people talked most frequently about two sources of commitment (in each case, 77 percent of the combined sample)—positive experiences of natural environments in childhood and adolescence, and family role models who demonstrated an attentive respect for the natural world. As in other studies, people gave several different explanations, including other experiences that reinforced these primary motives, but other reasons were mentioned notably less often (Chawla, 1999a). This work is descriptive research without comparison groups (Chawla, 1999b), but despite this reservation, the results of studies of this kind exhibit such strong recurring patterns that they merit attention.

Cognitive and psychoanalytic theories of development and social learning theory do not adequately account for the apparent importance of just *being in* and *appreciatively noticing* the natural world that these results suggest. Research in social learning theory demonstrates the importance of role models, expecially those who are nurturing, who control resources, and with whom a child identifies—characteristics of family members in well-functioning families (Bandura, 1986). This research fails to account, however, for the fact that the people I interviewed rarely mentioned explicit teaching or models of activism. More often, they mentioned family members who simply demonstrated how to be in nature with a secure attentiveness. For example, a woman biologist who had helped organize protests against the damming of rivers in Norway noted, "My mother knew the names of the plants more than other mothers did. So we talked more deeply about things. We didn't only fetch berries and fish, but talked about it" (quoted in Chawla, 1999a, p. 20). A lawyer in Kentucky who had helped lead the fight against the damming of the scenic Red River described how his father taught him how to make toys out of leaves and branches, find bait under rocks, quietly watch storms pass by, "and appreciate what's there" (ibid., p. 20). As often as family role models, people mentioned time in natural areas in childhood and adolescence as an independent motive in itself.

These memories of important childhood places accord with Gebser's description of archaic consciousness as sensory immersion and assimilation of the surrounding world, under effective conditions of well-being and security in which the child can peacefully and playfully be at one with its body and the world. Therefore, these experiences form a foundation for a wisdom that recognizes identity with the natural world as the origin of life, not just in evolutionary terms but, in Gebser's words, "ever-presently." The family members or other adults mentioned by the environmentalists could be described as "voices of appreciation" who encouraged the child to be in natural areas receptively, without barriers of inattention, fear, or defensive control.

Magic Relationships

What Gebser termed a magic experience of unity with the world, in its effective sense, was the focus of the work of Edith Cobb (1959, 1977). Based on a review of some 300 autobiographies that included childhood reminiscence, Cobb claimed that authors returned primarily to the middle years of childhood "to renew the power and impulse to create at its very source"—a source that she described as "a living sense of a dynamic relationship with the outer world" in which "the child appears to experience both a sense of discontinuity, an awareness of his own unique separateness and identity, and also a continuity, a renewal of relationship with nature as process" (1959, p. 539).

Intrigued by Cobb's claims, I reviewed her collection of autobiographies, which she had donated to the library of Columbia University's Teachers College. I found her selection of authors distinctive in several respects: almost all the authors grew up in Europe or North America in the nineteenth or early twentieth century and entered some field of the arts. To test her ideas further, I selected 38 recent autobiographies at random, and in this broader sample, I found that only authors who were in some way deeply involved in the arts or humanities described the kind of "magic" or "ecstatic" childhood experience of nature that Cobb discussed (Chawla, 1986). For most people, childhood memories of nature were primarily associated with family attachments. One-fifth of the authors, who came particularly from the physical sciences, politics, and business, described their childhood environments with detachment or rejection or omitted them altogether.

Perhaps magic experiences of nature are not universal, or perhaps only some adult temperaments and professions encourage their remembrance and recording. Nevertheless, 15 out of these 38 authors described experiences of this kind—attributing them to early childhood and adolescence as well as middle childhood (Chawla, 1990). Occasionally, people connected these memories to their impulse to create, but most frequently, they associated them with a fund of internal strength—like the British woman who recalled her experience on the moors. Some people believed that these experiences showed them the integration of nature and human life. In the words of the minister Howard Thurman, recalling how he walked an Atlantic beach by day and night as a boy: "I had the sense that all things, the sand, the sea, the stars, the night, and I were one lung through which all of life breathed. Not only was I aware of a vast rhythm enveloping all, but I was a part of it and it was a part of me" (quoted in Chawla, 1990, p. 21). When these experiences were reported, they were accorded a significance that Gebser's concept of magic union recognizes but that most developmental theories do not accommodate.

Many similar accounts appear in Hoffman's study of peak experiences in childhood. Hoffman (1992, p. 18) initially sought to interview children directly but quickly changed his method to a survey of more articulate adults about "childhood moments in which you seemed to experience a different kind of reality—perhaps involving a sense of rapture or great harmony?" In response to notices in newspapers and periodicals, he received more than 250 written and oral accounts, many of them about a suddenly transfixing sense of oneness with beautiful settings and elements of nature. Again, people most often attributed to this experience a fund of calm that they could later draw on.

Magic and Archaic Consciousness at Play

The preceding research relates to childhood indirectly, through memory. What relationships to nature do children themselves reveal? As preverbal states of being, archaic or magic consciousness cannot be spied by observation, and even less can we question children about them. In Hoffman's survey (1992), a frequent refrain at the end of people's accounts was, "At the time, I couldn't possibly have expressed what it meant to me" (p. 36). At their best, these ineffable forms of experience fall among what Little (1980) has categorized as children's "sweet nothings" (one of the reasons for the common dialogue between parent and child: "Where have you been?" "Out." "What did you do?" "Nothing.").

We can observe, nevertheless, that opportunities to be immersed in earth, water, and growing things and to make, destroy, and remake—mud pies, sand castles, rock piles, and such—are an important part of children's play outdoors (Tuan, 1978). Hart (1982) and Olwig (1986) have argued for the protection or creation of "wildlands" in all residential areas, where children can have this sense of absorption and power without trespassing on carefully managed adult domains. Recent evidence suggests that this absorption can be healing. In a study of 96 children with attention-deficit disorder, that related parents' ratings of their children's symptoms with independent ratings of the amount of natural elements in their play settings, children with greener play environments showed less severe symptoms in general, and children functioned better after activities in green settings (Taylor, Kuo & Sullivan, 2001).

We can also notice that children seek out natural areas for their play. Hart (1979), Moore (1980, 1986), and Sobel (1993) have documented this preference through extensive observations and interviews with children in North America, Britain, and the West Indies. A review of research on children's outdoor place preferences and place use shows a clear distinction between the streets, sidewalks, and yards near home, where children spend most of their observed time, and the natural areas and "wild spots" where they say they prefer to be (Chawla, 1992). In an international study of young adolescents' evaluations of their low-income urban communities, first carried out in the 1970s (Lynch, 1977) and replicated in the 1990s (Chawla, 2001), young participants repeatedly recommended the protection of existing trees and green places or the creation of more parks and gardens. Titman (1994) found similar results in children's evaluations of schoolyards, and Tapsell (1997) found that, almost without exception, children who lived near a channelized river recommended its restoration, seeing plantings, access to the water, and a return of wildlife as an increase in play value.

Surveys that have related children's environmental attitudes to their environmental experience also suggest the importance of time spent in natural places. Bunting and Cousins (1985) found that hiking, camping, and taking care of pets were associated with higher "pastoralism" scores among 1,100 nine to 17-year-olds, in addition to books and television shows about nature—*pastoralism* being defined as individual appreciation and concern for natural environments. Harvey (1989/1990) found that more schoolyard trees and nature centers, such as gardens and birdfeeders, were associated with higher scores for pastoralism among 845 junior school students.

In focus-group discussions with teenagers and adults in Singapore, which has a totally urbanized population, only individuals who had had opportunities to freely enjoy nature in their childhood expressed an intrinsic appreciation for natural areas, rather than fear or purely instrumental values (Kong, Yuen, Sodhi & Briffett, 1998). Similarly, Harrison, Limb, and Burgess (1987) found childhood experiences central to adult attachments to natural areas in a British city, and Kals, Schumacher, and Montada (1999) found that frequency of time spent in nature in childhood and experiences of nature with childhood family members significantly predicted a sense of emotional affinity with nature in a survey of 279 German citizens.

There is another category of children's "nothings"—"evasive nothings," Little (1980) has reminded us—that applies to dangerous and forbidden things that children have a deliberate reason to conceal. Through participant observation in children's play groups or through memories of childhood, several researchers have documented children's attraction for risk taking, often in natural areas (yes, especially boys) (Hart, 1979; Porteous, 1990; Chawla, 1994a). Risk taking invites a heightened awareness of the environment and a sense of power in overcoming danger that can be associated with Gebser's description of magic. When children are denied opportunities to experience sweet or evasive nothings, Little notes, they are left with "critical nothings"—a very real boredom and alienation because they do not have the resources they need. Little stresses that what is at issue is not just control over the environment—as if control were an end in itself—but opportunities for children to develop "a sense of perceived control over their environments that is based on an accurate reading of ecosystem constraints and resources" (Little, 1980, p. 12). In Gebser's terms, the world is full of examples of people's defective control over nature. What is needed is the development of a magic sense of power guided by wisdom that recognizes our real archaic identity with the natural world.

Mythic Places

As Gebser described mythic consciousness, it is an intuition of emotions and associative, metaphorical thinking that is known through the voice.

Therefore it is a world half created, half received through story, drama, song, and poetry, and a fundamentally social form of consciousness that communicates a sense of group experience and identity. Here nature figures as one of the most important protagonists in stories, as well as a metaphor for endless nuances of human feeling. On one end, there is the season of life "when yellow leaves, or none, or few, do hang / upon these boughs which shake against the cold" (in this case, as Shakespeare characterizes his old age in Sonnet 73) (1609/1977, lines 2–3). On the other, there are the countless boys and girls of fairy tale and fable who set forth into the forests of the world, often aided by the wild elements and creatures that they meet and often contending against them. How nature is presented in children's literature, film, and cyberspace and how children reconstruct these accounts in their own stories and art are rich areas for research (Engel, 1991; Rahn, 1995).

Stories invest the landscape itself with mythic significance. As Hester (1993) has observed, sacred spaces include not only famous places central to a culture's religious or national life but also the network of intimately known everyday places that define local identity for residents. In terms of the local landscape of children's experience, these places compose what Porteous (1990) calls "childscapes." This mythic world includes the landscape of children's dramatic play and all the landmarks that define the local worlds of any community of children—the swinging tree, the tree house, the monster dog, the lion in the tall grass. This way of experiencing nature has been extensively recorded through interviews and observations with children (Hart, 1979; Moore, 1986; Sobel, 1993). It includes the animated world that Piaget (1929) described, where the sun and moon follow young children on their walks and where stones need to be turned over so that they won't get tired of looking at the same view. Illusory as these mythic childscapes may seem from a rational perspective, they nevertheless represent a real world of feelings. In this storied world, it matters whether or not the moon is friendly . . . and if you are three feet tall, it will indeed appear wise to avoid the tall grass where lions may be lurking.

Children play "let's pretend" everywhere, and they animate houses and cars as well as sun and moon. Research suggests, however, that the quality of the environment affects the quality of play. In observations of play in different parts of a schoolyard, Kirkby (1989) found that children engaged in more dramatic play among bushes and other vegetation than on built equipment, and Moore and Wong (1997) reported decreased incidents of aggression and increased imaginative play and creative social interactions after a schoolyard had been converted from asphalt to an "environmental yard" with ponds, gardens, a meadow, and trees.

Another arena of mythic experience is community identity. Here, too, elements of nature affect the quality of children's lives. In international studies of 10- through 15-year-olds' evaluations of their low-income urban communities in the 1970s and 1990s (Chawla, 2001; Lynch, 1977), young people's priorities for good places in which to grow up were both social and physical: they wanted not only the provision of basic needs like water, sewerage, and shelter but friendly adults, places to meet friends, places with a range of interesting activities to observe or join, green places for play and gardening, and freedom from fear. A series of studies indicate that these qualities are not independent of each other. In a comparison of 64 public spaces in a low-income housing development in Chicago, Taylor, Wiley, Kuo, and Sullivan (1998) observed more than twice as many children playing in spaces with many trees than in those with few trees, significantly more creative forms of play in the spaces with many trees, and more access to adults' attention. Coley, Kuo, and Sullivan (1997) found that the presence of trees in inner-city neighborhoods predicted greater use of outdoor spaces by adults, youth, and groups of young and old together, and Kuo, Sullivan, Coley, and Brunson (1998) found that public housing residents who lived adjacent to public spaces with trees and grass reported more use of common spaces, more neighborhood social ties, and a greater sense of safety and adjustment.

Around the world, children live in increasingly urbanized societies (Chawla, 2001). If we accept the importance of archaic, magic, and mythic experiences, then how to provide children with access to positive experiences of nature within urban areas becomes a significant but not insurmountable challenge (Kellert, 1997; Chawla & Salvadori, in press).

The preceding research suggests that the preservation and restoration of trees and pockets of "nearby nature" in urban areas may go far in fulfilling the needs that children themselves express. What children say they want are both social and physical resources. Research with rural children shows that a universe of green, if it is barren of social and cultural opportunities, is no fair exchange (Matthews, Limb, Sherwood, Taylor & Tucker, 2000). To observe this is not to diminish the value of wild places, but not all children have the privilege of visiting them, whereas the preceding research suggests that all have a need for nature in their daily settings.

Conclusion: Toward Integral Experience

According to Gebser, every way of being makes a necessary contribution to the wisdom needed to solve our contemporary environmental crisis. He believed that human civilization now faces a crisis that is "planetary" or ecological as well as ethical and political and that the outcome is unclear (Gebser, 1985, pp. xxvii–xxviii). We may continue to be dominated by a defective mental consciousness that consumes human and ecological communities, or we may learn how to enter an integral consciousness that is open to the effective possibilities of the archaic, magic, mythic, and mental and that knows which form to use, on which occasions, "for the good of the whole"—a whole that, as Gebser has made clear, needs to involve "a new constellation of planetary extent"(ibid., p. xxvii). We may begin to move from a defective mental consciousness to integral consciousness, Gebser suggested, by shifting our attention from quantification to qualities of time.

In Gebser's words, this new consciousness is "not a freedom *from* previous time forms, since they are co-constituents of every one of us; it is to begin with a freedom *for* all time forms" (ibid., p. 289). By allowing an open play among all forms, we can distinguish the effective and defective possibilities of each. Like Wordsworth, who believed that the enduring influence of "spots of time," including childhood memories of nature, is an important resource for development in adulthood, Gebser believed that an integral wisdom requires that ways of being in the past remain accessible to the present.

In taking these positions, both Wordsworth and Gebser opposed the mechanistic and instrumentally rational view of nature that dominates modern consciousness, from the Romantic period to the present. Research on children and nature is not exempt from the dilemma of how to orient itself in relation to this view. How we, as authors of research and readers, relate to nature will influence what we see with regard to children in nature and what we neglect to see. Up to this time, research on this topic has been dominated by attempts to understand children's knowledge and attitudes about nature and their reasoning about environmental problems. These are important topics. Mental clarity about the complexity of problems is essential to finding solutions. Yet it is important to remember that this mental view represents only a partial perspective of our human experience of nature.

Gebser and Wordsworth draw our attention to ways in which our connection with the natural world depends on how we inhabit our bodies in the world. As Wordsworth argues that when a child receives love, sympathy, and care, it is prepared to reach out to the world around it with corresponding sympathy and creativity, so Gebser suggests that when we are grounded in the consciousness of our body with a sense of basic trust and security, we are enabled to accept identity with the world as the genesis of wisdom. These principles imply both that children need opportunities to identify with natural areas and that they, as well as the natural environment, need protection. In a world of diminishing natural areas and high rates of poverty and insecurity among the world's children, these principles point to the need for a new constellation of human relations with the world that must indeed be simultaneously ethical and political as well as ecological. Such a simultaneous effort is required to enable children to conceive bonds of attachment and identification with the natural world on the best possible terms.

Acknowledgment

I am indebted to Steven Holmes for emphasizing the importance of conditions of the body as well as qualities of the environment in determining children's experience of nature.

References

Abrams, M. H. (1971). *Natural supernaturalism.* New York: Norton.

Bandura, A. (1986). *Social foundations of thought and action.* Englewood Cliffs, NJ: Prentice-Hall.

Bruns, G. L. (1992). *Hermeneutics ancient and modern.* New Haven: Yale University Press.

Bunting, T. E., & Cousins, L. R. (1985). Environmental dispositions among school-age children. *Environment and Behavior, 17*(6), 725–768.

Chawla, L. (1986). The ecology of environmental memory. *Children's Environments Quarterly, 3*(4), 34–42.

Chawla, L. (1990). Ecstatic places. *Children's Environments Quarterly, 7*(4), 18–23.

Chawla, L. (1992). Childhood place attachments. In I. Altman & S. Low (Eds.), *Place attachment* (pp. 63–86). New York: Plenum Press.

Chawla, L. (1993, September). Earth origin. *Holistic Education Review,* 23–30.

Chawla, L. (1994a). Childhood's changing terrain. *Childhood, 2*(4), 221–233.

Chawla, L. (1994b). *In the first country of places: Nature, poetry and childhood memory.* New York: State University of New York Press.

Chawla, L. (1998). Significant life experiences revisited. *Journal of Environmental Education, 29*(3), 11–21.

Chawla, L. (1999a). Life paths into effective environmental action. *Journal of Environmental Education, 31*(1), 15–26.

Chawla, L. (1999b). Research methods to investigate significant life experiences. *Environmental Education Research, 4*(4), 383–397.

Chawla, L. (Ed.). (2001). *Growing up in an urbanizing world.* Paris/London: UNESCO/Earthscan.

Chawla, L., & Salvadori, I. (in press). Children for cities and cities for children. In A. Berkowitz, K. Hollweg & C. Nilon (Eds.), *Understanding urban ecosystems.* New York: Springer.

Cobb, E. (1959). The ecology of imagination in childhood. *Daedalus, 88,* 537–548.

Cobb, E. (1977). *The ecology of imagination in childhood.* New York: Columbia University Press.

Coley, R. L., Kuo, F. E., & Sullivan, W. C. (1997). Where does community grow? The social context created by nature in urban public housing. *Environment and Behavior, 29,* 468–492.

Engel, S. (1991). The world is a white blanket: Children write about nature. *Children's Environments Quarterly, 8*(2), 42–45.

Feuerstein, G. (1987). *Structures of consciousness.* Lower Lake, CA: Integral.

Feuerstein, G. (1989, May). Jean Gebser. *East West,* 62–95.

Gadamer, H. G. (1975). *Truth and method.* London: Sheed and Ward.

Gebser, J. (1962). *Anxiety.* New York: Dell.

Gebser, J. (1985). *The ever-present origin* (N. Barstad, Trans.). Athens: Ohio University Press.

Gould, S. J. (1977). *Ontogeny and phylogeny.* Cambridge, MA: Harvard University Press.

Guba, E. G. (Ed.). (1990). *The paradigm dialog.* Newbury Park, CA: Sage.

Harrison, C., Limb, M., & Burgess, J. (1987). Nature in the city: Popular values for a living world. *Journal of Environmental Management, 25,* 347–362.

Hart, R. (1979). *Children's experience of place.* New York: Irvington.

Hart, R. (1982). Wildlands for children. *Landschaft and Stadt, 14*(1), 34–39.

Harvey, M. R. (1989/1990). The relationship between children's experiences with vegetation on school grounds and their environmental attitudes. *Journal of Environmental Education, 21*(2), 9–15.

Heidegger, M. (1949). *Existence and being* (R. F. C. Hull & A. Crick, Trans.). Chicago: Henry Regnery.

Hester, R. T. (1993). Sacred structures and everyday life. In D. Seamon (Ed.), *Dwelling, seeing, and designing* (pp. 271–297). Albany: State University of New York Press.

Hoffman, E. (1992). *Visions of innocence.* Boston: Shambhala.

Holmes, S. J. (1999). Appendix C. *The young John Muir* (pp. 265–287). Madison: University of Wisconsin Press.

Hungerford, H. R., & Volk, T. L. (1990). Changing learner behavior through environmental education. *Journal of Environmental Education, 21*(3), 8–21.

Husserl, E. (1977). *Phenomenological psychology.* Trans. by J. Scanlon. The Hague: Martinus Nijhoff.

James, A., Jenks, C., & Prout, A. (1998). *Theorizing childhood.* New York: Teacher's College Press.

Kahn, P. H., Jr. (1999). *The human relationship with nature.* Cambridge, MA: MIT Press.

Kals, E., Schumacher, D., & Montada, L. (1999). Emotional affinity toward nature as a motivational basis to protect nature. *Environment and Behavior, 31*(2), 178–202.

Kellert, S. R. (1997). *Kinship to mastery.* Washington, DC: Island Press.

Kirkby, M. (1989). Nature as refuge in children's environments. *Children's Environments Quarterly, 6*(1), 7–12.

Kong, L., Yuen, B., Sodhi, N., & Briffett, C. (1998). The construction and experience of nature. *Tijdschrift voor Economische en Sociale Geografie, 90*(1), 3–16.

Kuo, F. E., Sullivan, W. C., Coley, R. L., & Brunson, L. (1998). Fertile ground for community. *American Journal of Community Psychology, 26*(6), 823–852.

Little, B. R. (1980). The social ecology of children's nothings. *Ekistics, 47*(281), 93–95.

Lynch, K. (Ed.). (1977). *Growing up in cities.* Cambridge, MA: MIT Press.

Matthews, H., Limb, M., Sherwood, K., Taylor, M., & Tucker, F. (2000). Growing up in the countryside. *Journal of Rural Studies, 16*, 141–153.

Mickunas, A. (1973). Civilizations as structures of consciousness. *Main Currents in Modern Thought, 29*(5), 179–185.

Moore, R. C. (1980). Collaborating with young people to assess their landscape values. *Ekistics, 47*(281), 128–135.

Moore, R. C. (1986). *Childhood's domain.* London: Croom Helm.

Moore, R., & Wong, H. (1997). *Natural learning.* Berkeley: MIG Communications.

Olwig, K. R. (1986). The childhood "deconstruction" of nature and the construction of "natural" housing environments for children. *Scandinavian Housing and Planning Research, 3*, 129–143.

Palmer, J., & Suggate, J. (1998). An overview of significant influences and formative experiences on the development of adults' environmental awareness in nine countries. *Environmental Education Research, 4*(4), 445–464.

Piaget, J. (1929). *The child's conception of the world.* New York: Harcourt Brace.

Porteous, J. D. (1990). *Landscapes of the mind.* Toronto: University of Toronto Press.

Rahn, S. (Ed.). (1995). Green worlds. *The Lion and the Unicorn, 19*(2).

Ricoeur, P. (1981). *Hermeneutics and the human sciences* (J. B. Thompson, Ed. & Trans.). Cambridge: Cambridge University Press.

Robinson, E. (1983). *The original vision.* New York: Seabury Press.

Searles, H. F. (1959). *The nonhuman environment.* New York: International Universities Press.

Shakespeare, W. (1977). *Shakespeare's sonnets* (S. Booth, Ed.). New Haven: Yale University Press. (Original work published 1609.)

Sobel, D. 1993. *Children's special places.* Tucson, AZ: Zephyr Press.

Soule, M. E., & Lease, G. (Eds.). (1995). *Reinventing nature?* Washington, DC: Island Press.

Tapsell, S. M. (1997). Rivers and river restoration: A child's-eye view. *Landscape Research, 22*(1), 45–65.

Taylor, A. F., Kuo, F. E., & Sullivan, W. C. (2001). Coping with ADD: The surprising connection to green play settings. *Environment and Behavior, 33*(1), 54–77.

Taylor, A. F., Wiley, A., Kuo, F. E., & Sullivan, W. C. (1998). Growing up in the inner city. *Environment and Behavior, 30*(1), 3–27.

Titman, W. (1994). *Special places, special people.* Crediton, Devon: Southgate.

Tuan, Y. F. (1978). Children and the natural environment. In I. Altman & J. Wohlwill (Eds.), *Children and the environment* (pp. 5–32). New York: Plenum Press.

Wals, A. E. J. (1994). *Pollution stinks! Young adolescents' perceptions of nature and environmental issues.* DeLier: Academic Book Center.

Wordsworth, W. (1952) *The poetical works of William Wordsworth* (E. DeSelincourt, Ed.). Oxford: Clarendon Press. (Original work published 1798.)

Wordsworth, W. (1971). *The prelude.* (J. C. Maxwell, Ed.) New Haven: Yale University Press. (Original work published 1850)

Zimmerman, L. K. (1996). Knowledge, affect, and the environment. *Journal of Environmental Education, 27*(3), 41–44.

9

Adolescents and the Natural Environment: A Time Out?

Rachel Kaplan and Stephen Kaplan

Our interest in adolescents and nature arose out of a perplexity. We have done considerable research on environmental preference and have been impressed by the consistency of the findings. With the exception of experts, people's preferences show a great deal in common across cultures, nations, and settings. One of the few studies to examine preference for different age groups, however, obtained results that were surprising given the widespread consistency. Adolescents appeared to show a distinctly different pattern from both younger and older groups. We wondered if this was a reliable finding: Do teens favor natural settings less than others do? Do they relate differently than others to natural settings? Do their environmental preferences reflect different patterns of needs and desires?

In this chapter we explore these questions. Our efforts to understand the relationship of teens and the natural environment lead to a far-ranging discussion of adolescence and the quest for meaning. We begin with the study that created the original perplexity and then examine other studies that used a comparable approach. We then turn to studies using different methodologies to see whether they provide further insights. Having explored the empirical literature, we conclude by attempting to place the findings in a larger perspective, both evolutionary and cultural.

Before we begin, however, some caveats: While the analysis here draws on some cross-cultural material, it is largely set in the context of youth in the United States in the late twentieth century. Observations made in an earlier part of the century or another part of the world might lead to some distinctly different conclusions. Furthermore, the concept of adolescence has itself seen dramatic changes both historically and cross-culturally; as the concept continues to evolve, some of our observations too could well become outdated before long.

One final caveat: This chapter is intended as a source of hypotheses or intriguing ideas; it is not intended to be a definitive treatise on this complex topic. Perhaps it can help shed light on some patterns in the research literature or lead to some innovative programs for youth that take advantage of their strengths and inclinations.

Preferred Environments

In general, people have a strong preference for natural environments. Substantial research in the last 30 years has documented this finding in different places, different cultures, and different populations, with the natural environments varying from the dramatic to the everyday (Kaplan, Kaplan & Ryan, 1998). These studies, based on ratings of scenes, filled an important gap in our understanding of people's feelings regarding nature. Prior to this research there was no empirical evidence concerning people's preference for different environments. This is not, however, all that these studies accomplished. They demonstrated that there is far more to preference than nature content alone. The additional insights make it clear that preference is not a casual or superficial matter but deeply grounded in human evolution and in the ways humans process and evaluate environmental information.

People view the environment as a source of information and decipher the information in terms of their potential competence in the setting (R. Kaplan, 1985). People like places where they can understand what is going on, can explore safely, and feel comfortable (Kaplan & Kaplan, 1995; S. Kaplan, 1992). If knowing one's environment is essential to survival, then favoring opportunities for exploration would be adaptive. At the same time, if one's survival depends on knowledge, then avoiding environments in which one had too little knowledge to function effectively would also be of great importance. Favoring settings that are understandable, therefore, has considerable adaptive value. Thus quite unexpectedly, the preference methodology has led to a new understanding of the relationship of preference and how people learn about their environment and function effectively in it.

In this section we review a number of studies that shed light on teens and nature preference, using ratings of scenes.

Developmental Study of Biome Preference

We alluded to a study, showing strikingly different preferences at midadolescence than for other age groups, that prompted our interest in teens and nature. Let us take a look at this study. Balling and Falk (1982) asked each of 548 participants to view 20 slides. The slides consisted of four examples of each of five biomes—savanna, temperate deciduous forest, coniferous forest, tropical rain forest, and desert. A panel of biologists and ecologists had rated a larger set of slides to help identify scenes that were most representative of each biome. The study participants were asked to rate each scene in terms of how much they would like to live permanently in such an area as well as how much they would like to see and visit such a setting. (The order of tasks was counterbalanced, and the scenes were shown in different random orders for the two tasks.) The study participants were drawn from diverse places and were "extremely mixed in terms of socioeconomic status, education, and the types of communities in which they lived" (ibid., p. 11). Particularly pertinent to our discussion is that the participants represented many age groups, including children in grades three, six, and nine (approximate ages of eight, 11, and 15, respectively). These children lived in Washington, D.C., or Anne Arundel County, Maryland. College students and older adults were also part of the sample.

As would be expected, across all age groups ratings were more positive under the "visit" instruction than for living in such settings, with the discrepancy greatest for the desert scenes. Across all groups, preference was greatest for visiting savanna settings, with the deciduous and coniferous forests not far behind. For the two youngest groups (ages eight and 11), preference for the savanna scenes was significantly higher than preference for any other biome, including the deciduous hardwoods, which would be far more familiar to these participants.

The most striking results, though receiving no discussion by Balling and Falk, were the notably lower preferences indicated by the participants in midadolescence (the 15-year-olds). For each of the biomes, the plots show a marked decline at this point, with the two younger age groups and the next two older age groups (college students and "35-year-olds") distinctly more favorable. These results could be interpreted to mean that midadolescent youths are simply more negative and likely to down-rate any scenes. What makes this interpretation less likely is that their responses to the different biomes showed strong differences. The desert scenes received far lower ratings than the rain-forest scenes and these were far lower than the savanna, deciduous, and coniferous forests. And while the latter three biomes were rated less favorably by the 15-years-olds than by other age groups, they all received mean ratings of around 4 on a six-point scale.

Other Photo-Based Studies

Australian Landscape The Australian landscape study (Herzog, Herbert, Kaplan & Crooks, 2000) is similar to Balling and Falk's (1982) with respect to permitting developmental comparison and examining natural environment scenes. While the Balling and Falk study included scenes from many places, this study examined the Bookmark Biosphere Reserve and its surrounding region in southeastern Australia. The 60 slides depicted rivers, dry lake beds (large open flat areas, generally devoid of vegetation), floodplains, terraces (gently sloping, generally treeless areas with a shrub layer for ground cover), mallee plains (sand ridge or dunes above the floodplains that are dominated by multistemmed eucalypt trees approximately 5 meters in height), and "cultural" scenes that included signs of either former habitation (such as sheep troughs) or planned agriculture. While Balling and Falk's participants were all from the United States, this study's participants were Australian children at two schools in upper-level primary (ages 10 to 12, n = 130) or secondary school (ages 13 to 17, n = 79), as well as college students and older adults. (American college students also participated in the study, but this aspect of the study is not pertinent here.) Participants were asked to indicate their preference for each scene using a five-point rating scale (5 = like it "very much").

Based only on the Australian sample (total n = 384), the age groups differed in overall preference [$F(3,135) = 39.56$, $p < .001$]. The primary students had the highest mean preference (3.22), the secondary students the lowest (2.79), with the other two groups in between and not different from each other (means of 2.98 and 3.00 for the adults and college students, respectively). The preference ratings were also submitted to factor analysis to permit comparison for different kinds of settings. Of the six resulting factors all but one showed the same pattern: the mean preference for the high school group was lower than either adjacent age group. The exception to the pattern was the factor consisting of five river scenes; here the two youngest age groups were very similar. For all age groups these rivers scenes were by far the most preferred.

Rural Forests Anderson (1978) examined visual preference for forest harvest and regeneration methods and land management practices in Lake County, Michigan. At the time of the study this rural county had high ethnic diversity, and forestry was a major factor in the economy, although unemployment was very high. The 300 study participants included resource professionals (n = 27), adult residents (n = 195), and high school students (n = 78) (equally divided among sophomores, juniors, and seniors), permitting comparisons based on expertise, ethnicity, and age. The visual preference component of the study involved 48 scenes presented as a photo questionnaire. Preference ratings of these scenes were used in a factor analysis that yielded five factors, representing different forest practices.

In terms of our discussion here, the noteworthy point is that there were no significant differences between the adult and high school age residents in their preferences for the five factors. The "planned spacious openings" were the most preferred scenes, and the "heavily manipulated landscapes" (clear cuts) were least preferred for these two groups. No data are presented that permit a finer-grain analysis in terms of younger or older teens.

Urban Context Medina's (1983) study was based on urban scenes, depicting housing, commercial uses, open space and recreation areas, and transportation. The participants included a national sample of 92 environmental educators, ages ranging from twenties to over 70, responding to a mail survey, and 207 youths, ages 12 to 14 (grades seven and eight), who were students in four predominantly African American Detroit schools. The two groups differed with respect to age, ethnicity, place of residence, and familiarity with the settings represented by the scenes.

Preference ratings by the two groups showed strong agreement with respect to the least liked scenes ("rundown urban" settings and "industrial/factory sites"). The preferred scenes, however, were quite different for the two groups. The teens liked scenes showing the kinds of neighborhoods in which they lived (single-family row housing and multiple-family housing) as well as scenes suggesting "urban mobility." The latter implied opportunities to explore and to bring the city within reach; thus even a scene showing a pair of city buses received a relatively favorable rating. By contrast, the educators expressed greater appreciation for nature settings, albeit nature in a very urban context. Scenes with tree-lined streets, enclosed open spaces, and low-density housing were among their most preferred. The older sample tended to prefer scenes suggesting privacy and quiet, whereas the students had higher preference for settings that suggested activity, places where one could do things.

River Corridor A photo questionnaire was used as one source of public input in developing a master plan for an urban corridor paralleling a river (Kaplan, 1989). It was distributed to a randomly drawn sample of city residents, property owners, and employees within the study corridor and to anyone who requested a copy. In addition to these 506 study participants, 115 high school students, enrolled in social studies class at three local high schools, completed the survey. There is no information about their age or grade level (tenth through twelfth grades). For purposes of the discussion here, the high school students are compared to the random sample ($n = 128$) as they are most comparable in not representing any particular stake in the outcome of the study.

The 40 scenes comprising the photo questionnaire (five pages, eight scenes per page) were selected to represent a wide variety of possible alternative treatments for the study area, including different kinds and degrees of commercial development as well as more or less developed natural areas. Participants rated each scene in terms of how much they would like it if it were in the study area, using a five-point scale (5 = very much).

Factor analysis of the preference ratings yielded three factors. The "light industry" settings were consistently least preferred, and the random sample and high school students were similar in their low ratings. Scenes comprising the "urban development" factor depicted a highly built, urbanized context. Preferences for these scenes ranged broadly, but the high school students' ratings for the factor were significantly more positive than the random sample's (means 2.8 and 2.3, respectively, $t = 4.88$, $p < .001$). Figure 9.1 (top right and bottom right) shows the two scenes the teens rated most highly in this set. Their mean rating of 3.6 for these compares to the other group's means of 2.3 (top right) and 1.9 (bottom right). These dramatic differences are not unlike the different preferences found by Medina (1983).

Scenes comprising the third factor, "nature and wooden walkways," were the most preferred by both the random sample and high school students. The two groups differed significantly, however, in how much they liked them (means 4.2 and 3.7, respectively, $t = 4.43$, $p < .001$). Eight scenes received ratings above 4.0 for the adult sample, while only two were in this category for the high school students. One of these (figure 9.1, bottom left) was the most preferred scene for both groups. The other "nature" scene in figure 9.1 (top left) and the two urban scenes received equivalent ratings (mean 3.6) by the teens, while for the adults this nature setting merited a mean of 4.1.

Summary These studies have in common that they ask participants to view visual images (slides or photographs) and rate them individually in terms of how much they like them. The environments represented by these scenes are different in terms of many dimensions, including their homogeneity and the extent of being built or natural settings (for example, Balling and Falk's set included only natural settings, and the River Corridor study ranged widely). Furthermore, the settings differ with respect to their familiarity to the participants. The ages of the teen samples also varied considerably. Medina's participants were in grade seven or eight, Balling and Falk's in grade nine, and participants in the other two American studies were in grades 10 through 12, while the Australian secondary school students included a broader age band (13 through 17).

Despite the differences in settings and developmental stage, the results of these studies have considerable similarity. They strongly suggest that adolescents, compared to younger and older groups, have lower

Figure 9.1
Kaplan's (1989) River-Corridor Photo Questionnaire. The lower left scene was rated highest in preference by the high school students (mean 4.2) as well as the adult random sample (mean 4.7). For the teens the other three scenes were equivalent in preference (mean 3.6). The adults, by contrast, favored the other nature

Figure 9.1 (continued)
scene (upper left, mean 4.1) and had much lower ratings for the two scenes on the right from the "urban development" factor (means of 2.3 and 1.9 for top and bottom, respectively).

preference for natural settings and greater appreciation of certain kinds of developed areas. The latter tend to be places that suggest action and activity. The results do not show that youths do not appreciate nature; where comparison is possible, the nature settings tend to receive higher ratings than the other scenes. Ease of access also does not account for the teens' preferences. In the river corridor study, for example, the highly preferred urban settings were not accessible to these participants, while many of the nature settings were readily available to them. Nor do the studies suggest that teens use the five-point rating scale differently or that they simply tend to dislike everything. In the river corridor study there was about a half scale point difference between the teens and the adults, with the direction of difference depending on the content of the photograph.

Other Indications of Adolescents' Environmental Preferences

Visual images are a useful way to represent and sample environments. Other aspects of environmental preferences are better captured through other data-gathering techniques. In this section we look at some of these indications of adolescents and their preferred settings. In some cases we draw on additional data obtained in studies mentioned earlier; these permit comparison across age groups. In other cases, only adolescents were included in the study.

River Corridor In addition to the photographs discussed earlier, this survey included verbal items describing potential changes in the uses and activities in the study area. Participants were asked to rate each of these in terms of their desirability. Factor analysis of these ratings yielded four factors. For two of them, "residential/office" and "small business" uses, the teens and adults were equivalent and unenthusiastic (means around 2.0). The most desirable uses for both samples were included in the "parks and recreation" factor, although the mean was significantly higher for the random sample than the high school students (4.1 and 3.7, respectively, $t = 2.76$, $p < .01$). The groups were particularly discrepant in their views of the desirability of nature trails and a nature center. For these items the teens' endorsement was substantially lower than the adults'. "Fishing pier" was seen as least desirable by both groups. Similar ratings were given to jogging trail, bike path, and riverfront urban park.

The remaining factor based on the desirability ratings was labeled "festival center," with means for the two groups around midscale (3.0 and 2.8 for the teens and random sample, respectively, $t = 1.92$, $p < .05$). Of the seven items comprising this factor two ("boutiques, festival market" and "restaurants and cafés") received similar ratings by the two groups. The other items showed contrary perspectives. The adults were more favorable than the youths about inclusion of "museum / cultural center" and "band shell / amphitheater." The high school students, by contrast, were far more enthusiastic than the random sample participants about a "conference facility / hotel," a "fountain in the river," and especially about "shopping center / mall." The last of these was for the adults one of the least desirable uses of the riverfront area. Viewed as a whole, the items in this factor that the teens indicated as most desirable all suggested places or activities where teens can hang out and be part of the action. Vanderbeck and Johnson (2000) report similar preferences in their study of inner-city youth.

Favorite Places Several studies have asked adolescents to indicate their favorite places. These have shown remarkable consistency across many countries. In her study of Sunshine, Australia, Owens (1994) found that teenagers prefer more developed, less natural places. When she asked 101 teens (ages 13 through 19, though mostly 14- and 15-year-olds) to name three outdoor places they valued within the community, the most frequently selected places were "developed parks" (38 percent) where they typically went for recreational activities including football, soccer, and tennis. The next most often selected were "places at home" such as their own or a friend's backyard (17 percent), and "commercial areas" (17 percent) including the corner stores, downtown pedestrian shopping areas, and mall parking lots. Almost all the youths (91 percent) indicated that they bring others to the place or go because others are there (63 percent).

Lynch (1977) reported very similar patterns in a UNESCO-sponsored study of early adolescents in Australia, Poland, Mexico, and Argentina. In answer to questions about "where they like best to be, where they feel most at ease, where it is best to meet with friends and to be alone," the youths in Melbourne consistently replied that their own room, home, or homes of friends were top on their list. In Salta, Argentina, the pattern was much the same, with the addition of the plaza and local street corners (ibid., p. 48). Polish adolescents' top choices were their own home and "green area," which included park, sports field, or meadow (ibid., p. 138). Green places were also a frequent choice among Mexican youths.

Korpela (1992) reported comparable results in Finland, where 248 adolescents (ages 17 and 18) most often cited private homes as their favorite places (39 percent of the sample); bars, cafés and discos, sports facilities, and natural settings were mentioned with equal frequencies (about 15 percent). By contrast, Korpela (1991) found that 63 percent of an adult sample in the same city mentioned natural settings as their favorite places.

A Time in the Wilderness The studies described thus far suggest that the teens have a strong inclination for places that depend on the urban infrastructure. At the same time, however, they appreciate the natural environment. The Outdoor Challenge Program we were involved with in the 1970s and early 1980s (R. Kaplan, 1984; Kaplan & Kaplan, 1995) included both teens and adults in a large wilderness area in and around the McCormick Experimental Forest in Michigan's Upper Peninsula. During the final two years of this program (1980 and 1981), the outings lasted nine days and included a 48-hour solo period during which participants spent time at a lakeside location, out of sight of anyone else. The days before the solo involved relatively arduous hiking through dense areas, swampy regions, and some steep country, much of it without trails.

Participants knew that the program was part of a research project; this meant that their costs were largely subsidized and that they would be asked to complete questionnaires at various times. The eight groups that participated in the program during these two years included 22 adults (17 women and five men, ages 19 through 48) and 27 high school-age youths (12 boys and 15 girls). Most of the youths came from schools in the Upper Peninsula; some of the adults also came from this region, while others were residents of southeastern Michigan. Participation in a program of this kind involves a strong self-selection factor; individuals seeking this kind of adventure are likely to prefer natural settings and to have had experience with camping and backpacking. Nonetheless, very few of them had been on a solo previously.

On being picked up from their solo sites, participants were asked to complete a questionnaire about their experience. Items for these questionnaires were based on open-ended and structured data collected during prior years of the program. They included questions about their explorations during solo, their reflections, and their feelings about being alone. The results showed some interesting contrasts between the adults and the youths. The experience of being alone was far more positive for the older participants. They were more likely to find it a "great experience," to enjoy the peacefulness, and to enjoy "just being alone." The adolescents, by contrast, were significantly more likely to find it both lonely and boring and to experience the "silence" as a less positive time (means on cluster of items were 4.4 and 3.5 for the adults and youths, respectively). The adults also indicated that they did more exploration of their surroundings during their solo, getting to know "a lot about a small area" and "the mysteries of nature."

Even though these youths opted to participate in the Outdoor Challenge Program and were aware that the program included a solo, they clearly found this aspect of the experience more trying and less wholesome than other parts of the program. That is not to say that they regretted doing it. In fact, in hindsight the entire Outdoor Challenge experience was transforming for many of the participants (Kaplan & Kaplan, 1995).

The Urban Hypothesis While the youths on solo in the Outdoor Challenge Program found the experience less peaceful and more lonely than the adults, they nonetheless were intensely involved in a natural setting for an extended period of time. There is no doubt that many of their age mates would not choose to participate in such a program; there is certainly an assumption that urban youths would be less comfortable in nature settings. Lewis (1982) found this to be true. He described taking urban teenagers to an arboretum and contending with their queries about the presence of lions, tigers, bears, and snakes. The study by Bixler, Carlisle, Hammitt, and Floyd (1994) also supports this assumption.

However, since the youths in the Outdoor Challenge Program were generally from rural backgrounds, they may not have had such fears. Bixler and Floyd's (1997) study is useful in including a large rural group (62 percent of the 450 participants) as well as suburban (22 percent) and urban (15 percent) youths—all in eighth grade in schools in Texas. Data were collected using structured questionnaires in the classroom setting. They found that youths who scored high in disgust related to nature, worries or fears about the natural environment, and desire for indoor comforts were more likely to prefer manicured park settings and urban environments and to dislike wildland (unmanicured) environments. Fear expectancy was a particularly strong positive predictor of preferences for manicured park paths and negative predictor of wildland job preferences. Fear expectancy included such concerns as getting lost, "stepping on a snake," "getting separated from friends," and "being chased by a swarm of bees." As the authors conclude, the pattern of results among a sample of predominantly rural and suburban youth contrasts with the typical assumption that "it is urbanites who tend to respond negatively to natural environments" (ibid., p. 461).

Nearby Natural Beauty Whether the youths are from rural or urban settings, the fear of nature might reflect lack of direct experience (Bixler et al., 1994). Both Owens (1988) and Hester et al. (1988) reported that adolescents who live nearby natural beauty do appreciate it. Participants in Owens's study of teens living near Mt. Diablo State Park indicated that "to be with nature" was a strong reason for their preference for the natural park. In the Hester et al. study, 90 percent of the teens listed the cliffs overlooking the ocean as the most special place in town. Nonetheless, these adolescents expressed some ambivalence; while they appreciate the beautiful, peaceful, and safe natural places in their community, they long to be where the action is or to have more action come to them.

Teen Leisure Patterns One way to be where the action is involves participation in formal organizations. In their study of over 3,000 rural eighth-graders, Willits, Crider, and Funk (1988) found that 80 percent of those surveyed belonged to one or more school-sponsored clubs or organizations, though the youths who lived more than 15 minutes from school were less likely to participate in school activities (especially interscholastic sports teams). Youth participation in formal organizations outside of school, however, was limited. About one-third participated in some kind of church-sponsored activity, and 75 percent reported no community participation other than school or church. In fact, they report that by far the most common leisure-time pursuit was socializing with friends. Second and third were socializing with family and watching TV.

This pattern of rural youths is not substantially different from the results Obasanjo (1998) found in his study of the effect the physical environment on inner-city youths. While the study focused on the impact on cognitive functioning of housing and neighborhood quality as well as opportunities for restorative experiences, the data that are particularly pertinent to our discussion here involve the adolescents' leisure time activities. The study included 660 inner-city teens, with between 140 and 184 in each of four groups (ages 14, 15, 16, and 17, respectively). These four age groups differed significantly with respect to several activities: "socializing, or hanging out with friends" was popular for all ages, but the youngest group rated it 4.2 (on a five-point scale), while the mean frequency for the 17-year-olds was 3.9. The age differences between the 14- and 17-years-olds were even more dramatic for sports (means 3.5 and 3.0, respectively) and watching television (means 4.1 and 3.4, respectively).

Larson (2000) provides a similar picture in his review of adolescents' leisure patterns. He reports that 40 to 50 percent of adolescents' waking hours during the school year are devoted to leisure and that the amount increases in the summer. About 7 to 14 percent of this time is devoted to watching TV. Further, "A large proportion of Western adolescents' leisure time is spent in activities with peers, mainly talking and hanging out" (ibid., p. 173). Larson further points out that these activities involve a high degree of intrinsic motivation, yet little concerted effort. School work, by contrast, entails the opposite pattern.

Little's (1987, 1998) analysis is particularly useful in showing that the social patterns, while perhaps low in "concentration," nonetheless have substantial meaning. He found that teens most often reported their current personal projects to focus on interpersonal, academic, or recreational activities. He further asked his participants to rate their projects in terms of a variety of attributes. The interpersonal and academic projects were found to be equivalent in ratings of their "importance." They differed dramatically, however, in how meaningful and enjoyable they were to the youths. The academic projects "are seen as important duties of significant value. However, they are projects in which adolescents see little of themselves being expressed and feel precious little in the way of mild pleasure, let alone delight" (1998, p. 201). The highly meaningful projects involved "intimacy or connectedness."

Summary Although diverse in context and methodology, these studies point to some common themes. Adolescents appreciate natural settings, though apparently not as much as do younger children or adults. They favor places where they can be with their peers and activities that convey excitement and action. To the extent that natural settings support these inclinations, they are preferred.

The Adolescent Agenda

Scanning the titles of publications under the heading of "adolescence" one could readily reach the conclusion that this is the name of an illness or some combination of misfortunes. Owens (1997) documents the awesome increase in material on teens in the popular literature that reflects "social problems" (from 90 articles to 420 in a 30-year period) or "substance abuse" (16 to 244 in the same period). By contrast, articles reflecting "activities/hobbies/recreation" showed little change. There is no question that the nature of the teen years has changed in the last quarter century; however, it had also shown many changes in the decades before that. A brief historical view is useful for an analysis of current patterns and for thoughts for the future.

Urbanization and poverty are pertinent themes in this context. At the same time, however, attributions to urban (versus rural) setting and low-income may be only partially appropriate. Ladd's (1982) perceptive analysis based on her work with black youths provides an important example. She wrote: "When we consider where young people of previous generations in the U.S. sought and found adventure, our thoughts turn to natural settings, wildlife, and open spaces. Vanishing are the natural areas, especially wooded areas, in and around cities, where, only a few decades ago, city kids explored, charted, roamed, hid, were lost, and, the lucky ones, found safe and unhurt. . . . There is no place—no natural environment—left for the urban adolescent to explore and experience adventure" (ibid., pp. 444–445). She suggested that some less sanctioned activities of urban youth living in poor areas (stealing cars, shoplifting, pulling fire alarms) can be explained by this loss of opportunities for "legitimate adventure."

Ladd's advocacy for safe settings for approved excitement and adventure for urban youths still holds today. However, it is less clear that the availability of nearby wooded areas and open spaces would solve the problems to which she alluded, either in the 1970s or currently. Anderson's (1978) work was conducted at about the same time as Ladd's. The African Amercian youths in his study lived in a rural area where woods and open spaces were far more available than shopping malls. Yet they did not seek the woods for adventure. Why were such places attractive at an earlier time and less so in more recent decades? Have the woods changed, or are the teens different?

Changing Trends in Adolescence

As it turns out, the nature of adolescence has changed in many important ways. To be sure, Socrates and Aristotle expressed displeasure at the behavior of their adolescent students (Harris, 1998; Obasanjo, 1998), suggesting that some of these patterns are of long standing. However, there are also indications that youths who were expected to be responsible members of society (for example, where their help on the farm was imperative) exhibited different patterns. Kett (1977, p. 3) provides a fascinating analysis of some of the historical transformations and their far-reaching implications:

> Compared to their predecessors in 1800 or 1900, young people in the 1970s spend much more time in school, much less at work. They are essentially consumers rather than producers. Their contacts with adults are likely to occur in highly controlled environments such as the classroom, and the adults

encountered are usually conveyors of specialized services such as education or guidance. For the most part, young people in the 1970s spend their time in the company of other young people.

Kett traces some of the changes to the transformation from a largely agrarian society to an increasingly urbanized pattern. This had profound demographic implications, including reduction in family size, narrowing of age ranges among siblings, and, in time, different age distributions in the population. These patterns later showed further dramatic changes as a consequence of the baby boom (births between the end of World War II and the early 1960s). Schulenberg and Ebata (1994) cite an increase in the proportion of 14- to 24-year-olds from 15 percent in 1960 to 21 percent in 1975 with a drop back to 16 percent fifteen years later.

Leaving the farm had further ramifications with respect to patterns of livelihood and the educational support necessary for these new careers. The Depression provided an additional impetus toward increased education given the lack of jobs. In the first half of the twentieth century the proportion of teens graduating from high school showed astounding increases: 6 percent in 1900, 51 percent in 1940, and 62 percent in 1956 (Kett, 1977). The pattern of extending education has become dominant not only in the United States but in most highly developed countries, making the educational context the defining characteristic of "adolescence" (Hurrelmann, 1994). In fact, Schneider and Stevenson (1999) attribute much of the steep rise in adolescents' "ambition" to greater educational aspirations of today's teens, with the vast majority of American high school seniors (90 percent, they say) expecting to attend college.

Other changes occurred after World War II. Affluence rose sharply as did the number of cars in the garages of the newly built homes in the American suburbs. Many more 16-year-olds obtained driver's licenses as the legal age for driving dropped in many states and a "driver's license became the rite of passage for many middle-class American young people" (Kett, 1977, p. 265). Greater affluence also fed enormous increases in consumptive goods marketed to the teen population and increasingly promoted through television. These included clothes, jewelry, cosmetics, musical records, and entertainment opportunities.

Concurrent with these changes were dramatic drops in participation by youths over age 14 in voluntary youth-serving organizations such as scouting, Y-sponsored programs, and 4-H clubs. The strong adult influence in the conduct of these organizations has been a major cause of their declining membership. Already by the 1970s, "domineering adult leaders" (Lipsitz, 1977) was mentioned as a prime cause of disengagement by adolescent girls who had previously participated in such organizations.

During this era there has also been a sharp increase in two-wage-earner families and in single-parent households. The norm of children living with a mother and a father and only the father working outside the home has shifted markedly, with fewer than one-fourth of families fitting this pattern by the late 1980s (Schulenberg & Ebata, 1994). While most children still live in two-parent households, that number has declined substantially, and for children who live with a single parent, the rate of poverty is significantly higher. Both of these patterns (two wage earners and single parent) translate to less parental time with adolescent children. Increasingly, youths have turned to each other for social support and have spent much of their time in a subculture with a very narrow age range.

Many factors have converged to make adolescence a distinctly different stage in life now than it had been historically. Perhaps the most important impacts of these changes has been the strong reliance on peers as the major support system.

Culture and Evolution

The shift in emphasis for the teen years and the increasing importance of peers for this age group point to important cultural factors. A cultural framework need not, however, be viewed as an alternative to an evolutionary interpretation. We fully agree with Midgley's (1978) succinct statement to the effect that humans are a species that has evolved to need culture. Among her many illustrations of interactions between evolution and culture is the example of fast locomotion, a topic that is certainly apt for teens. People everywhere are interested in fast locomotion, but as she points out, the form this takes varies widely from culture to culture.

Our exploration of the possible interaction between evolution and culture focuses on the theme of human needs and particularly needs that are informational. In much of our previous work on cognition and environmental preference (Kaplan & Kaplan, 1982; Kaplan et al., 1998) we have discussed the human dependence on information and the consequent needs to seek it, understand it, and act on it. Considerable evidence suggests that humans evolved to be particularly adroit information processors and that their survival depended on this capability (Lee & Devore, 1968; Pfeiffer, 1978). The particular ways they fulfill these informational needs, however, are likely to vary over the course of the life span. Furthermore, the patterns for accomplishing these needs have changed dramatically over the course of human history and will, no doubt, continue to change. The motivations to fulfill these needs are likely to have strong biological roots and reflect cultural factors in their expression.

Informational Needs

We focus on three needs with far-reaching implications. Two of them, the need to explore or seek information and the need for understanding, are closely linked. New information that is difficult to integrate with old information readily leads to confusion, which is aversive. People often show a preference for gathering new information at their own pace to expand and enhance what they already know and comprehend. Another important aspect of exploration is the degree of risk. This parameter appears to vary developmentally. The child who is just learning to walk is at the same time shifting from a relatively fearless exploration of the surrounding world to one that is far more constrained by fear (Marks, 1987). This developmental shift is, of course, adaptive; the possibility of encountering danger is far greater when one can locomote efficiently, allowing one to escape the watchful eye of the protective parent. The toddler, in other words, is at the developmental state in which the risk parameter of exploration is turned down to a low level.

The adolescent explores at the opposite end of the risk spectrum. The timing of this pattern of more risky exploration seems appropriate for multiple reasons. This is an age of comparatively high strength and agility, along with relatively lower level of responsibility to others than will, before long, be the case. The increased risk taking is seen in trying things where one has little prior experience, in going places one has little knowledge of, and in attempting to do things that are acknowledged to be difficult, thereby demonstrating that one is no longer a child.

The third need we consider involves taking action. The motivation to act in an effective way, although relying on understanding and exploration, is perhaps an even more basic human need. From an evolutionary point of view, a complex and powerful brain is of no consequence unless it can lead to adaptive action. Further, this contribution to action must be sufficient to offset some of the hazards of a more complex brain, such as the possibility that one will become lost in thought and the greater susceptibility to emotion (Hebb & Thompson, 1954). For these reasons, there must be a bias to act and to act in ways that are adaptive for the individual. This means such action must have strong motivational support and be an important source of satisfaction. Clearly acting to secure resources or otherwise foster one's goals would constitute action of this kind. But humans are, as we have seen, highly social animals. Thus one would also expect actions that make one more valued by one's groups would be adaptive. In particular, actions that demonstrate competence and actions that show helpfulness to others would improve one's standing in the group. This collection of adaptive actions, related to resources, one's goals, and the requirements of one's group, we propose to call "meaningful action."

Meaningful actions particularly salient to the adolescent years are likely to include ones that speak to the needs for both autonomy and social support. Desires for "autonomy and self-determination . . . self-focus and self-consciousness, [and] salience of identity issues" (Eccles et al., 1993, p. 94) are strong themes of the adolescent years. Comparably Magen (1998, p. 47) speaks of the adolescents' needs for independence and for freedom to make their own decisions, while pointing out that they also show "strong need for guidance, relationship, and, not infrequently, dependence." The focus on self-interest or autonomy, on the one hand, and the need to be socially connected and even socially useful, on the other, are hardly unique to teens. Goldschmidt (1990) pointed out a similar duality in his broadly based anthropological treatise, *The Human Career: The Self in the Symbolic World.* He cites a vast array of cross-cultural data to support the idea that people organize their lives to gain the respect of their fellows. This notion of "careers" focuses on "the trajectory through life which each person undergoes, the activities he or she engages in to satisfy physical needs and wants and the even more important social needs and wants" (ibid., p. 106). Although he does not argue the point, the emphasis on gaining respect of the members of one's group makes good evolutionary sense. And as one sheds the protective armor of being "just a child," it becomes adaptive to win the respect of one's peers as well as the community at large.

Although the increased focus on autonomy and the quest for social support can be attributed to cultural factors, our hypothesis is that they have evolutionary bases as well.[1] Autonomy and independence are also characteristics of adolescents in other species. Schaller's (1964) observations of gorillas in their natural setting provide insight into the possible agenda of individuals approaching adulthood. He described a situation in which a group of adolescent males appeared intent on taking over the leadership of the troop. While probably stronger than the older males in charge, they were not well enough organized to achieve their goal. Ultimately they left the troop, taking with them other young members.

This observation is interesting in a number of respects. Separation of a group into smaller units when it reaches a certain size may in itself be a useful adaptive strategy (Price & Stevens, 1999). In addition, the inclination of the young to be dissatisfied with where they were raised provides a mechanism for the species to explore other possible locations as well as alternative patterns. Spear (2000), who characterizes these changes as independence and risk taking, points out that they occur in other species as well. She further bolsters her position by suggesting that the extensive "remodeling" of the brain that occurs in adolescence may account for these characteristic behavioral changes. The needs for exploration, understanding, and taking action are thus strongly expressed in the adolescent's focus on autonomy and quest for social support. The manifestation of these expressions is firmly based in one's contemporary culture. Thus in cultures that offer extensive guidance and encouragement for youth to explore adult roles, the peer group may be relatively less influential. The contemporary American culture, however, offers little guidance in identifying challenging tasks that are meaningful. The media-carried messages, by contrast, support a culture of consumerism and a laissez-faire mandate of "do your own thing." In this context the inclination of the American adolescent to look for meaningful action at the mall is hardly surprising. It is, however, only part of the story.

Despite much popular press to the contrary, there are plenty of examples that show that adolescents are often not selfish, passive, and unconcerned individuals; they often care deeply to make a difference. Little (1998, p. 200), for example, mentioned that adolescents who were engaged in "community volunteer projects appraised these as exceptionally self-expressive." Magen's (1998) work directly addresses the question of how adolescents find personal fulfillment and what impact social commitment has on their happiness. She summarizes the results of studies carried out over an extended period of time and spanning different cultures as demonstrating "a powerful relationship between the intensity of young people's happy moments and their readiness to commit themselves to others" (ibid., p. 161).

There are indeed many examples of teens who are involved in their communities and in design decisions as well (Owens, 1997). Chawla (1997) documents a variety of community action projects involving adolescents in many parts of the world. Her book *Growing Up in an Urbanizing World* (in press) documents the ways youth have taken part in planning as well as implementing improvement to their urban setting. Mullahey, Susskind, and Checkoway (1999) offer examples of youth participation in community planning efforts in a number of cities.

While the teen years are a time of many contradictions and personal change, they are not necessarily different from other stages in the life cycle in terms of underlying needs and desires. We have conceptualized informational needs that are basic to human functioning as the needs to make sense, to explore, and to take meaningful action. These needs are modulated both by developmental factors that foster preparation for autonomy and by cultural factors that guide this preparatory activity. The developmental influence expresses itself both in a change in focus and in heightened riskiness in the service of exploration. Culture, in turn, provides guidance in terms of what being an autonomous adult requires. To the extent that guidance of this kind is lacking, the adolescent, already inclined to question authority as part of the shift to autonomy, looks to the peer culture to provide the missing information.

The "Time Out" Revisited

Learning about the physical environment is a lifelong concern for members of our species. Survival once depended on knowledge of the environment; to some degree it still does. Preference is a mechanism (like many other mechanisms, partly innate and partly learned) that pulls the individual toward some environments and not toward others. Thus, from an evolutionary perspective one of the adaptive functions of preference is to guide exploration and hence to foster learning. The possibility of acquiring more information is itself a powerful attraction and an important element in preference.

As we have seen, however, various studies in different settings and using distinct methodologies suggest that there may be a temporary decrease in this inclination to be attracted by the physical environment. This phenomenon appears to occur during the period when children begin to prepare to take on adult roles. During that time, the priorities of children in Western culture seem to shift away from the physical environment toward other concerns. Many of these do not focus so much on place as self and peers. Furthermore, there is greater attention to one's own role and competence. Youth at this stage might thus be expected to be attracted to environments that afford opportunities for independent action, for testing alternative patterns, and for gaining respect from peers and/or the community.

From an adaptive perspective the shift in preferences to give more weight to the work of transition to adulthood should be temporary. Once the transition is well under way, the substantial preference for the natural environment would be expected to return. This is the case for several reasons. For the major part of human history the natural environment was the basis of making a living. In evolutionary times, when technology was little developed, knowledge of the environment was particularly important (Laughlin, 1968). And since the environment is ever changing, continued contact is necessary if one's knowledge is to stay up to date. There are also the demands of the new identity of the individual; taking on the responsibilities of adulthood would be likely to require new and more differentiated knowledge of the environment. Hunters, for example, need a great deal of knowledge about potential prey, to say nothing of the importance of being able to find their way home. Gatherers need to know a great deal about what they gather, such as the distinguishing characteristics of edible and nonedible plants, their likely location, and the ideal time of harvest (Flannery, 1955). Thus there is a great deal of new information to be acquired from the environment once the adolescent begins to prepare for adult responsibilities.

In principle, the shift to a lowered nature preference in adolescents could be explained in terms of either culture or evolution. What is particularly striking, however, is that the phenomenon is of relatively brief duration. There are no obvious cultural cues that seem to dictate the return to a strong preference for nature. One might argue that the shift occurs once the individual has achieved autonomy. In fact, however, the shift seems to come at the age when autonomy is probably achieved in cultures that are less committed to prolonging formal education. The typical college freshperson is a long way from autonomy. Nonetheless, a number of studies have shown high nature preference for this group (e.g., Kaplan, Kaplan & Wendt, 1972). Thus the return from the "time out" appears to be more likely the result of a developmental, and hence evolutionary, influence.

While the very "differentness" of this age group is often a source of frustration, recognizing the needs that are particularly prominent in the teen years may provide clues for how to structure programs for this time-out interval. We have seen that this group has a strong social orientation, with a particular emphasis on their peers. They also have need for self-determination—for exercising autonomy and making choices. And they have a concern for acquiring and displaying competence, both in terms of strength and skill. Nature-related activities that meet these needs are likely to qualify as meaningful action in the eyes of the adolescent.

Many nature-related activities could be tailored to meet these needs. To mention a few examples, youth can be engaged in ecological restoration projects, creating a boardwalk across a dune, helping design a community open space that meets their needs, or teaching younger children in a nature setting. The success of these efforts, however, is likely to depend on a few key factors. Awareness of youths' sensitivity to autonomy, social concerns, and competence needs are certainly important, but they are not enough. It is essential to take the time and effort to find out "where they're at"—what activities in nature would be perceived as meaningful and satisfying to the potential participants. Such respect for the teens' own insights and inclinations is essential for identifying what would constitute meaningful action.

A related yet distinct factor addresses adolescents' sensitivity as to who sets the rules and who initiates the activity. If activities are perceived as adult-generated (as is true for much of schooling), they have a much reduced likelihood of gaining favor. By contrast, "responsive and demanding" orientations to parenting have been found more likely to lead to positive outcomes (Collins et al., 2000); the same issues are applicable to programs for teens. The "responsive" component suggests listening and hence participation, while the "demanding" dimension points to the central role of challenge and responsibility. Hart's (1997) adaptation of the Arnstein ladder of citizen participation provides imagery about how youth can be engaged not only peripherally but even in initiating ideas and sharing in the decision-making process.

Conclusion

We began our quest with the perplexity created by Balling and Falk's finding of lower preference for nature scenes on the part of adolescents. We feel that the evidence, scattered as it is, does support the original suspicion that there is a "time out" in preference for natural environments during the adolescent years. This does not mean that adolescents dislike nature but rather that nature settings do not hold the powerful pull for teens that they do for those younger or older. While for many teens there is some discomfort with nature places, there is no indication that they would avoid contact with nature if it were the context for activities that effectively meet their needs. Our analysis led to identifying some of these needs and to suggesting ways that programs and activities could be structured to foster them.

Fortunately, many of our suggestions are integral features of some programs and activities that are designed for teens and build on the strengths of this age group. At the same time, however, the people who have keen insights about these issues are often not people who do research about them. As a result, a great deal of wisdom and empirical support about what works, for whom, and under what circumstances is not available for those who have not arrived at these useful intuitions. We have no way of knowing whether such teen-oriented activities result in increased preference for nature settings for this age group. We strongly suspect, however, that participants in such activities gain familiarity and comfort with the natural environment, which serve them well as the time-out years give way to adulthood.

Acknowledgments

We are grateful for the Cooperative Agreements with the U.S. Forest Service, North Central Forest Experiment Station, Urban Forestry Project, that have funded different aspects of the work discussed here. We want to thank Olusegun Obasanjo for permitting us to include an analysis based on unreported portions of his doctoral dissertation. Nancy Wells provided substantial help in identifying and tracking down relevant literature. We also want to thank Robert Bixler, Louise Chawla, Myron Floyd, and Patsy Owens for their very welcome input. We greatly appreciate the helpful suggestions provided by the editors, Peter Kahn and Stephen Kellert.

Note

1. Innately influenced developmental patterns are evident at many stages in the life cycle. Infants display a strong focus on learning about objects and how they behave (Mackworth, 1976). The disappearance of fearlessness that occurs at about the time children begin to walk is another vivid example of developmental expressions of evolution (Buss, 1999). Comparably, the adolescent skepticism about whether the adults running the show know what they are doing turns out to have a striking counterpart in the adolescent gorilla (Schaller, 1964).

References

Anderson, E. (1978). Visual resource assessment: Local perceptions of familiar natural environments. Doctoral dissertation, University of Michigan.

Balling, J. D., & Falk, J. H. (1982). Development of visual preference for natural environments. *Environment and Behavior, 14*, 5–28.

Bixler, R. D., Carlisle, C. L., Hammitt, W. E., & Floyd, M. F. (1994). Observed fears and discomforts among urban students on school field trips to wildland areas. *Journal of Environmental Education, 26*, 24–33.

Bixler, R. D., & Floyd, M. F. (1997). Nature is scary, disgusting, and uncomfortable. *Environment and Behavior, 29*, 443–467.

Buss, D. M. (1999). Evolutionary psychology: A new paradigm for psychological science. In D. H. Rosen & M. C. Luebbert (Eds.), *Evolution of the psyche* (pp. 1–33). Wesport, CT: Praeger.

Chawla, L. (1997). Growing up in cities: A report on research underway. *Environment and Urbanization, 9*(2), 247–251.

Chawla, L. (Ed.). (in press). *Growing up in an urbanizing world*. London: Earthscan & Unesco.

Collins, W. A., Maccoby, E. E., Steinberg, L., Hetherington, E. M., & Bornstein, M. H. (2000). Contemporary research on parenting: The case for nature and nurture. *American Psychologist, 55*, 218–232.

Eccles, J. S., Midgley, C., Wigfield, A., Buchanan, C. M., Reuman, D., Flanagan, C., & MacIver, D. (1993). Development during adolescence: The impact of stage-environment fit on young adolescents' experiences in schools and in families. *American Psychologist, 48*, 90–101.

Flannery, K. V. (1955). The ecology of early food production in Mesopotamia. *Science, 147*, 1247–1256.

Goldschmidt, W. (1990). *The human career: The self in the symbolic world*. Cambridge, MA: Blackwell.

Harris, J. R. (1998). *The nurture assumption*. New York: Free Press.

Hart, R. (1997). *Children's participation: The theory and practice of involving young citizens in community development and environmental care*. London: Earthscan.

Hebb, D. O., & Thompson, W. R. (1954). The social significance of animal studies. In G. Lindsey (Ed.), *Handbook of social psychology*, Vol. 1, Theory and Method. Cambridge, MA: Addison-Wesley.

Herzog, T. R., Herbert, E. J., Kaplan, R., & Crooks, C. L. (2000). Cultural and developmental comparisons of landscape perceptions and preferences. *Environment and Behavior, 32*, 323–346.

Hester, R. T., Jr., McNally, M. J., Hales, S. S., Lancaster, M., Hester, N., & Lancaster, M. (1988). "We'd like to tell you . . .": Children's views of life in Westport, California. *Small Town, 18*(4), 19–24.

Hurrelmann, K. (Ed.). (1994). *International handbook of adolescence.* Westport, CT: Greenwood.

Kaplan, R. (1984). Wilderness perception and psychological benefits: An analysis of a continuing program. *Leisure Sciences, 6*, 271–290.

Kaplan, R. (1985). The analysis of perception via preference: A strategy for studying how the environment is experienced. *Landscape Planning, 12*, 161–176.

Kaplan, R. (1989). The tension between development and open space: Insights from public participation. In G. Hardie, R. Moore & H. Sanoff (Eds.), *Changing paradigms* (pp. 193–198). Oklahoma City, Environmental Design Research Association.

Kaplan, R., & Kaplan, S. (1995). *The experience of nature: A psychological perspective.* Ann Arbor, MI: Ulrich's. (Original work published 1989)

Kaplan, R., Kaplan, S., & Ryan, R. L. (1998). *With people in mind: Design and management of everyday nature.* Washington, DC: Island Press.

Kaplan, S. (1992). Environmental preference in a knowledge-seeking knowledge-using organism. In J. H. Barkow, L. Cosmides & J. Tooby (Eds.), *The adaptive mind* (pp. 535–552). New York: Oxford University Press.

Kaplan, S., & Kaplan, R. (Eds.). (1982). *Humanscape: Environments for people.* Ann Arbor, MI: Ulrich. (Original work published 1978).

Kaplan, S., Kaplan, R., & Wendt, J. S. (1972). Rated preference and complexity for natural and urban visual material. *Perception and Psychophysics, 12*, 354–356.

Kett, J. F. (1977). *Rites of passage: Adolescence in America 1790 to present.* New York: Basic Books.

Korpela, K. (1991). Adolescents' and adults' favourite places. In T. Niit, M. Raudsepp & K. Liik (Eds.), *Environment and social development. Proceedings of the East-West Colloquium in Environmental Psychology, Estonia.* Tallinn: Tallinn Pedagogical Institute. Pp. 76–83.

Korpela, K. (1992). Adolescents' favourite places and environmental self-regulation. *Journal of Environmental Psychology, 12*(3), 249–258.

Ladd, F. (1982). City kids in the absence of legitimate adventure. In S. Kaplan & R. Kaplan (Eds.), *Humanscape: Environments for people* (pp. 443–447). Ann Arbor, MI: Ulrich. (Originally published 1978)

Larson, R. W. (2000). Toward a psychology of positive youth development. *American Psychologist, 55*, 170–183.

Laughlin, W. S. (1968). An integrating biobehavior system and its evolutionary importance. In R. B. Lee & I. DeVore (Eds.), *Man the hunter.* Chicago: Aldine.

Lee, R. B., & DeVore, I. (Eds.). (1968). *Man the hunter.* Chicago: Aldine.

Lewis, C. A. (1982). Nature city. In S. Kaplan & R. Kaplan (Eds.), *Humanscape: Environments for people* (pp. 448–453). Ann Arbor, MI: Ulrich. (Originally published 1978.)

Lipsitz, J. (1977). *Growing up forgotten.* New York: Heath.

Little, B. R. (1987). Personal projects and fuzzy selves: Aspects of self-identity in adolescence. In T. Honess & K. Yardley (Eds.), *Self and identity: Perspectives across the lifespan* (pp. 230–245). London: Routledge & Kegan.

Little, B. R. (1998). Personal project pursuit: Dimensions and dynamics of personal meaning. In P. T. P. Wong & P. Fry (Eds.), *The human quest for meaning: A handbook of theory, research and applications* (pp. 193–212). Mahwah, NJ: Erlbaum.

Lynch, K. (Ed.). (1977). *Growing up in cities: Studies of spatial environment of adolescence in Cracow, Melbourne, Mexico City, Salta, Toluca, and Warszawa.* Cambridge, MA: MIT Press.

Mackworth, J. F. (1976). Development of attention. In V. Hamilton & M. D. Vernon (Eds.), *The development of cognitive processes.* New York: Academic.

Magen, Z. (1998). *Exploring adolescent happiness: Commitment, purpose and fulfillment.* Thousand Oaks, CA: Sage.

Marks, I. M. (1987). *Fears, phobias, and rituals.* New York: Oxford.

Medina, A. Q. (1983). A visual assessment of children's and environmental educators' urban residential reference patterns. Doctoral dissertation, University of Michigan.

Midgley, M. (1978). *Beast and man: The roots of human nature.* Ithaca, NY: Cornell University Press.

Mullahey, R., Susskind, Y., & Checkoway, B. (1999). Youth participation in community planning. *Planning Advisory Service Report No. 486.* Washington, DC: American Planning Association.

Obasanjo, O. O. (1998). The impact of the physical environment on adolescents in the inner city. Doctoral dissertation, University of Michigan.

Owens, P. E. (1988). Natural landscapes, gathering places, and prospect refuges: Characteristics of outdoor places valued by teens. *Children's Environments Quarterly, 5*(2), 17–24.

Owens, P. E. (1994). Teen places in Sunshine, Australia: Then and now. *Children's Environments, 11*(4), 292–299.

Owens, P. E. (1997). Adolescence and the cultural landscape: Public policy, design decisions, and popular press reporting. *Landscape and Urban Planning, 39*, 153–166.

Pfeiffer, J. E. (1978). *The emergence of man.* New York: Harper and Row.

Price, J., & Stevens, A. (1999). An evolutionary approach to psychiatric disorders: Group-splitting and schizophrenia. In D. H. Rosen & M. C. Luebbert (Eds.), *Evolution of the psyche* (pp. 196–207). Westport, CT: Praeger.

Schaller, G. B. (1964). *The year of the gorilla.* New York: Ballantine.

Schneider, B., & Stevenson, D. (1999). The ambitious generation: America's teenagers, motivated but directionless. New Haven, CT: Yale University Press.

Schulenberg, J., & Ebata, A. T. (1994). The United States. In K. Hurrelmann (Ed.), *International handbook of adolescence.* Westport, CT: Greenwood.

Spear, L. P. (2000). Neurobehavioral changes in adolescence. *Current Directions in Psychologocical Science, 9*, 111–114.

Vanderbeck, R. M., & Johnson, J. H. (2000). "That's the only place where you can hang out": Urban young people and the space of the mall. *Urban Geography, 21*, 5–25.

Willits, F. K., Crider, D. M., & Funk, R. B. (1988). Small-town youth related their recreational preferences. *Small Town, 18*(4), 14–18.

10

Adolescents and Ecological Identity: Attending to Wild Nature

Cynthia Thomashow

A change is taking place, some painful growth, as in a snake during the shedding of its skin dull, irritable, without appetite dragging about the stale shreds of a former life, near blinded by the old dead scale on the new eye. It is difficult to adjust because I do not know who is adjusting: I am no longer that old person and not yet the new.

—*Snow Leopard*, Peter Matthiessen (1978)

Three years ago I walked into the Reptile House at Chicago's Brookfield Zoo and shuddered. I came upon a group of six teenagers transfixed on a python molting its skin. The process of molting has always fascinated me. Watching a snake wriggle out of a transparent yet binding outer shell is exhausting, as you can feel the effort it takes to slowly free the supple new skin from its drier, brittle casing. Like Peter Matthiessen (1978) above, I find this process of "shedding of skin" a strong metaphor for personal transformation. I wondered how these teens would react.

The teens cheered the python on while simultaneously shivering and expressing repulsion at its behavior. They seemed unaware of their collected movements, which mimicked the undulations of the snake. I watched a group dance of sinuous and raw sensuality. After 20 minutes of intense concentration, one of the members began to mention a likeness between the python's molting behavior and changes in human personality. Each one of the teens told a story of "shedding" undesired traits and commented on the difficult process of change and the vulnerability attached to it. Moreover, one teen said: "I felt such an intense shiver of knowing when I looked into the eyes of the snake it made my skin crawl, as if I knew everything at once, like the spirit of the world lay behind those yellow eyes and he was beckoning me out." The snake's action seemed to confirm deep instinctual knowledge that a force much larger than themselves held a script for evolution, a key to understanding the process of growth and change.

In this chapter, I want us to traverse the edges of the adolescent world to discover how the tensions of adolescence play themselves out in the milieu of the natural world. I believe there is a powerful connection to "wild nature" in the teenage years that can be tapped and utilized both developmentally and educationally.

I begin by reviewing the stresses and challenges of adolescence that seem at home the context of nature. I discuss the adolescent connection to nature under the rubric of ecological identity. Then I describe three school-based programs that have integrated ecological thinking into the educational experience of teens. The first engaged high school students in the management of public lands within a small New England city. The second engaged high school students in the protection of a wildlife sanctuary in rural New Hampshire. The third engaged eighth-grade students in the design of an exhibit for the Brookfield Zoo. My goal here is to show how each program helped foster adolescent ecological identity and attended to the essential wild nature of adolescent development.

The World of the Adolescent

The episode in the Reptile House is one of many in which I have witnessed teenagers allude to a strong affiliation with nature. The snake stimulated a powerful discussion about adolescent growth, change, and transformation. When the lines of distinction between nature and the self disappeared, the teens entered into a way of thinking about themselves through nature. This process is referred to by Paul Shepard (1996, p. xiv) as ecological thinking: "Ecological thinking requires a kind of vision across boundaries. The epidermis of the skin is ecologically like a pond surface or a forest soil, not a shell so much as a delicate interpenetration."

What is particularly exciting about drawing on the construct of ecological thinking is that adolescence is a critical time of identity formation. I am interested in linking teenagers' thoughts about nature to the development of a coherent self-image, creating an ecologically grounded identity. Teens are embedded in the process of piecing together a lifetime of values, beliefs, experiences, and behaviors. Elements are continually added and discarded during these years. Deep thought mingles with superficial whim, intense concentration mixes with daydreaming, resistance melts into acceptance. Teens gather, sift, and choose the elements of human culture and wild nature that will be woven into an adult identity that will, hopefully, be coherent, pragmatic, and ethical.

Working with young people requires a discerning eye and the ability to see beneath the surface. Because adolescents often relish forms of expression that rouse adult reaction, we tend to disregard what might be important signals of change and growth. Adolescents often provoke by first changing their appearance, which then allows them to adopt new behavior. One disguise is easily exchanged for another as they test the waters to find the persona that matches their new beliefs and values. Such provocations can draw our attention, cause confusion, and create distance, or they can act as a signal to look more closely. If educators understand adolescent development, they can more easily see through to the root of "wild" behavior and resist the temptation to run.

With interest and sometimes verve adults can attend to the adolescent behind the dramatic persona. Here is an example of "wildness" as it emerged one afternoon in the Boston subway. The young man I encountered was what many would classify as a "punk." His hair stood 10 inches high, in a Mohawk-style cut, dyed bright red. He wore a dog collar with spikes and lots of black leather with a breechcloth-style apron hanging front and back filled with different aboriginal symbols. His eyebrows were pierced, as were his tongue, nose, and lips in various places. He initially rejected my overtures for conversation with growls that were meant to scare. But with good humor I persisted, and eventually we settled into an interesting conversation. He described the unbearable wildness that seemed to scratch at him, compelling his rowdy behavior and unusual appearance. He spoke of his remorse for the numbness that seems to take its place in adulthood. He spoke of his need to physically distinguish himself from adults who he perceived were "selling out." His costume expressed individuality and a value stance; it told a story of identity unfolding and emerging.

The prowess of this age lies in the willingness of adolescents to be contentious and argue the big issues of the world. Their interest is in forming arguments, ripping the insides out of ideas and rebuilding them based on current beliefs and values just to see if they will last through a discussion. It is essential for teens to compare and contrast, to quarrel and compromise, to emote and appease, to rationalize and defend out loud the ideas they are considering keeping.

Schools that expect one "right" answer are forcing a lid on valuable synergistic energy. Experimentation and combinatorial variation, metaphorical magic, and hypothetical permutations reign supreme in the frontal lobes of this group. As educators, we must be prepared and willing to tap the possibilities.

Paul Shepard (1996, p. 5) writes:

> Adolescence is a preparation for ambiguity, a realm of penumbral shadows. Its language includes a widening sensitivity to pun and poetry. Appropriate to its psychology is attention to the zones between categories, zones that have their own animals. The borders from which obscenity and taboo arise are figured in creatures that embody a sense of overlapping reality: the insects that crawl between two surfaces, the owl flying at dusk, the bat that seems to be both bird and mammal. The adolescent person is a marginal being between stages of life, on the shifting sands of an uncertain identity. In this respect his symbols are changeling species: the self-renewing, skin-shedding snake, the amphibious frog that loses a tail and grows legs, the caterpillar that metamorphoses into a butterfly.

While it is important not to overreact to the mutating outer shell and the numerous adolescent "sheddings of skin," it is important to use this shape shifting as a sign that something inside is changing and to take note. The flux and tumult of this age offer an opportunity for adults to help sculpt the final product. Educators can become influential mentors in this process when they are willing to brave the current.

Adolescents face powerful sexual impulses. They seek autonomy but also intimacy. They are defined by their relationships and steeped in self-consciousness. Their increased sensitivity to perspective taking brings up issues of respect, reciprocity, equity, and fairness as they struggle to work these concepts into their ethical decisions and actions. They are seeking to make meaning of the world and to find their place in it. And from a still-emerging ethical stance, adolescents are seeking to transform the world.

It is not my goal to provide a full account of these characteristics; however, I do want to add the often overlooked ecological component of adolescent development. My central thesis here is that nature can—and I think must—play a key role in the healthy development of adolescent identity.

Ecological Identity as a Model for Teaching Adolescents

Nature is often used by adolescents as a template to better understand and manage the tensions of adolescence. As Paul Shepard (1996) noted, profound images can be drawn from natural phenomena that aid in the unraveling of developmental mysteries. I would like to share several examples of writing by adolescents that refer to nature as a means of reflecting on growth and transformation.

Through my student teachers and as a consulting practitioner, I have been able over the years to gather a large quantity of written works and oral narratives of middle and high school students. I have also kept a journal of "close encounters" with adolescents (such as the experience in the Reptile House). I use these moments of revelation and reflection as signals of deeper meaning making. They identify ecological thinking and an affiliation with nature that are deeply connected to their struggle to define themselves.

The first poem was written by a young man who was considered outgoing and popular while in high school. It exemplifies the fight to guard an inner self that feels too fragile for public display and an ache for the relief that nature provides from the pressure to conform and perform:

Any Other Day
It felt routine, like any other day
Of trudging through this apathetic school,
Detached from friends, not meaning what I say,
Befogged in clouds I laugh at what is cruel
Until such grayness clears to show the fool
Is me, and then like lightning I am struck
And sudden loneliness as stark and cool
As night swamps me, I cry, mired in muck
And yet in this morass I am not stuck,
For in the clear and daunting dark I find
My heart, to know myself, the one who snuck
Away from this life, to accept in kind
Some unconditioned sense of who I am,

Then darkness yields to light which flows undammed.
To see me in this place of no people, no scam
Only naked before what is out there unmanned.
—Mark, age 17

Through nature, adolescents are privy to models of living other than the cosmetically driven social world of magazines and movies and to rhythms and cycles that are different from those imposed by the constructs of a school day. Through nature they gain access to the wild and untethered, the naked realities of life and death, and the basics of survival and come face to face with their biological origins and the underpinnings of human purpose and meaning.

Adolescents often claim a landscape, an animal, or some other natural feature as a metaphor for growth and development. It might be a place they know or a specific object that serves metaphorically to reflect a place of personal meaning, a mood, or an inner struggle, as exemplified in the piece below:

The Island
The rocks have texture and style
The whine of a gull is the soundtrack
And the sun sets the mood
Cliffs surround me and leave a circle on my neck
I am obsessed with observing
Granite builds like blocks hundreds of feet up
Grass and spruce cling to the side like lichen
Smell, sight, and sound have flooded my thoughts
The island has me and I have the island
—Jake, age 15

Finding or constructing a "home place" in nature can serve as the incubator for the developing self. David Sobel (1993, p. 23) comments on the necessity of both constructed and found places in the lives of children: "I suspect that it is the sense of self, the ego about to be born, that is sheltered in these private places. The onset of puberty in adolescence initiates an often painful focus on 'Who am I?' The construction of private places is one of the ways that children physically and symbolically prepare themselves for this significant transition."

In my high school teaching, I engage students in an activity called a "sense of place map" (see Thomashow, 1995, p. 192). Students select a place that they have come to love or identify with during their lives—

one that they refer to or think about frequently and to which they feel a strong connection. Very often, students will speak about places they have "lost" for various reasons. One student wrote about the wooded lot in back of her house that was sold for commercial and residential development:

> My parents betrayed me by selling this piece of land. I used to wander there, spend time thinking and sorting out things. I loved playing back there. As the bulldozers moved in, I remember feeling like my arms and legs were being torn from my body, I felt it deep down inside myself, like I lost a part of myself. It was awful and sad and unreconcilable. How could they do such a thing without consulting me? I will never forgive them. (Tracey, age 18)

This deep sense of loss uncovers a profound love for the natural world in which Tracey found solace, a place to think and separate from the confusion and tumult of everyday life. She bonded so thoroughly with the "wooded" area that having it leveled felt like a personal violation. Her essay went on to describe the "activism" that spawned from this loss. Tracey and her friends tried to sabotage the bulldozers one night, leaving red "go away" signs on their yellow flanks. The event has fueled other protests, as well. Tracey often organizes her peers to march against suburban sprawl and attend town meetings about rezoning land for commercial development.

Nature provides the solid infrastructure in a world of swirling possibility, a place to return for stability and balance, a place that unguardedly provides the real stuff of life. Concrete experiences in nature contribute to a resource bank of material to draw on in the construction of the self, fulfilling what seems to be an intuitive understanding of the need to include one's ecological self in the mix of "Who am I?" As Paul Shepard (1996, p. 5) writes:

> For human beings, habitat and environment are the literal space of the ground of thought. What the child wants is to find a place in which to make a world the way the world is made. The home range of the ten-year-old is the first context of spatial and temporal thought. The child is a "traveler" mapping out the first spatially ordered reality of his life. The end of childhood is the end of that simple identity. The literal fauna have become the external expression of the child's own congeries of feelings and bodily processes, a community of self-confidence.

The recognition and understanding of our visceral connection to nature have the potential to shift the way we conceptualize the world and how it works, shaping an ecologically minded sense of purpose and responsibility in the way we behave. To call on Paul Shepard (1996, p. xv) again, "When the self is expanded to encompass the world, environmental destruction becomes self-destruction." If left to atrophy, this aspect of identity seems to shrivel and recede to the dark reaches of our consciousness. Unfortunately, the majority of adults live out their lives in dull awareness of their connection to nature, never clearly determining its influence on the way they see and consider the world. My interest is in finding ways to capitalize on the raw "wild nature" of adolescence and tap into the ecological connections that seem so near the surface, vibrantly linking ecological thinking to the creation of identity.

Programmatic Approaches to Ecological Identity

Mitchell Thomashow (1995) outlines a comprehensive educational program that is successful in reconstructing the ecological identity of adults. I borrow heavily from his work when designing educational programs for high school students. In brief, ecological identity "refers to how people perceive themselves in reference to nature, as living and breathing beings connected to the rhythms of the earth, the biogeochemical cycles, the grand and complex diversity of ecological systems" (ibid., p. xiii). The approach is easily integrated into high school curricula by encouraging teachers to use the direct experience of nature as a framework for the personal decisions, professional choices, political action, and spiritual inquiry of adolescents.

Social and cultural norms, expectations, and rules are putty in the hands of adolescents. The current ecological paradigm—the worldview that directs our relationship to nature—is fair game. For example, the concept of "nature as a resource for human consumption" is eagerly dissected, revealing grand questions about how humans are connected to nature. These meaty issues are the grist of discussions one might have in a classroom that integrates ecological thinking into traditional subject matter, and they rise organically from the problems that confront communities in which these teens live.

The educational equation that has worked best for me is a melding of ecological identity work that considers the unique character of adolescents and focuses on real community issues. I would like to describe three different school-based experiences that consciously integrated the essential character of teens into a study of the local environmental issues. These experiences use nature as an integrating context for learning and as a focus for building ecological identity. They have a deliberate interest in exploring questions about relationship to nature by immersing teens in relevant local environmental dilemmas.

Adolescents and the Management of Public Land

Last year I walked into a meeting that I never dreamed would actually take place. City council members lined the wall facing two rows of young people from the local high school and middle school. They were discussing the management of a local piece of land. A striking element in this meeting was the difficulty adults seemed to have in listening to the students. Every time a high school student started to talk, the mayor or the city planner cut them off. Tension was rising. Bo Hoppin, the manager of this school-based project, was a strong advocate for the kids, but he was letting them feel their way through this resistance.

The students had important data that related to the management of this land and a clear idea of appropriate land use. They had analyzed the "worth" of this 2.3-acre plot for a year, surveying for "natural resources," testing water quality, collecting soil samples, counting populations of wildlife, and monitoring human use. The information showed that this land was in recovery from a long history of human use, including the overflow of a mill pond, a dump for solid waste, logging, and agriculture. The kids had been promised complete discretion in determining the future use of this plot. Now that the decision was on the line, the adults were getting cold feet. To renege now would damage the fragile trust of these young people. Bo believed in them, and they knew it. Feeling his strength, they plowed through the heavy doubt of the city leadership.

Three years ago I was invited on a hike to assess the future of a 2.3 acre section of land abutting a larger city park. The 297-acre Ashuelot River Park was a well-kept secret in Keene, New Hampshire, until a small group of citizens created a public gateway by converting a parking lot into a manicured riverfront area. The 2.3-acre piece sitting off to the side of the new entry had been relatively untouched for about 100 years. I jumped at the prospect of giving this small portion of the public park to the youth of the city as a learning laboratory and field research station.

There is a lot of groundwork to making a project like this one work. Multiple players—including the school system, the city government, the citizen group of overseers, high school students, and staff—had to agree on the approach, the goals, and the outcome. When Bo Hoppin took over the management of this project, our intent was to stay true to our promise that teens would control what happened to this piece of land.

We started by enlisting an Advanced Placement environmental science class at the high school, actively engaging students in the collection of base-line date. Students used the land and water to hone their research skills. Working with their teacher and a university professor, they performed natural resource inventories, soil investigations, water-quality testing, and historical analyses of the property. Each student developed an independent research project that would help develop a profile of the land. They presented their findings to the city planning board. That is where the real work started! Getting adults to respect and believe in the integrity of their work brought politics and economics into the picture.

I knew that the students were highly invested in the outcome of this meeting. Students sat for hours unobtrusively observing wildlife, drawing impressions, and recording data. They braved cold and wet, committed long hours, and creatively dealt with technical and mechanical difficulties, willingly giving up big chunks of their leisure time to sort and classify, record, and synthesize relevant data. They believed that they were doing meaningful work. This was the real thing: they were solving real problems and making an important contribution to the place where they lived.

Sitting solitary for hours in the forest made an impression that can't be quantified or statistically analyzed. Many kept research journals with margins spilling notes about awe-struck observations of wildlife, pieces of poetry, or inspired articulations of their own reflection in water, birds, or trees. One student wrote:

In the night I dream, and imagine where I am, of how I came to be, and how there came to be land. What is water, sky, and space? What is rain, sleet, and snow? I have so many questions, some answers I'd like to know. These questions fill my head, when I am close to sleep, in my dreams I am inspired to find the answers that I seek. Dreaming lets me travel through water, space, and sky and create the answers to my questions, while in bed I lie. (Jenna, age 17)

Slowing down the pace of their lives and immersing themselves in the sounds and movement of birds, small mammals, and insects had a significant impact on the way these teens began to think about themselves. They started to comment on their relationship to the weather, other species, the run of the river, and the quietness of this place as if self was reflected in the sanctity of this land, and it fueled their passion to protect its integrity. In protecting this piece of land, they were protecting a piece of themselves.

The students decided that people should be able to circumnavigate this 2.3-acre area but not go into the interior. The interior section was restricted to the fauna and flora that called it home. They garnered the participation of elementary school students to design and plant an indigenous garden that would attract wildlife and restore the native flora. Middle-school students created exhibits that informed the public of the restoration plans and gave natural history information to visitors. A kiosk was designed and built by middle-school children that stands at the perimeter, warning people against intruding on the sanctuary and explaining why.

The controversy at the City Council meeting centered on a grant that was received by the planning board (without consulting the kids) to build a boardwalk into the interior of the 2.3 acres allowing greater access and providing an observation deck. The students wanted to guard against overuse and had decided against building the boardwalk because of its impact on wildlife and vegetation.

Emotions grew hot as the students held their ground. Finally, the student spokesperson stood up. "This is not about the boardwalk," she said with her voice shaking. "This is about including us as citizens in this decision and keeping your promise. We didn't take this responsibility lightly. We have done our best to research and make the best possible decision for the good of the land and the people of Keene. You simply have to believe in us and honor your commitment. This is about whether you really think we are a part of this city, whether we deserve to make a decision that effects the place where we live. Your decision to involve us, or not, will determine our future commitment to this town and to public decision-making. It will effect whether we think we can really make a difference." This articulate and emotional statement broke the back of the resistance. It created the foundation of the program that continues today. High school students have been asked to manage seven other pieces of public land in Keene.

Students like Hannah Jacobs, who spoke out at the meeting, are now in college combining the study of political science and environmental studies. When the Rachel Marshall Outdoor Learning Laboratory (RMOLL) was officially dedicated as part of the larger city park, Hannah gave the opening speech (edited below):

> I think the most incredible aspect of the laboratory is its level of community and student involvement. Many students are scared and upset when they see the natural landscape changing around them. We felt, for a long time, like we had no voice in the development of our community. RMOLL actually gave us a chance to make a positive impact and have our voices heard. We were participating in the preservation of a piece of land in our town and deciding how that land would be managed in the future.
>
> A concept from a book is never truly understood until it is seen as part of the natural world. Here, students can touch the trees, plants, soil, and water. They can search for animal tracks, sample the river, and see for themselves how a dead tree provides valuable habitat for wildlife. When I was doing research here for my biology project, I stood in awe as I watched a Cooper's hawk hunting chickadees for a noon-day snack. As well as providing a beautiful place for research and learning, it is students' connection to the real world. It is real because we helped plan this piece of land. It is real because we now see our plan in action and the many purposes it serves. (Hannah Jacobs, age 17)

Adolescents and the Protection of a Wildlife Sanctuary

About three years ago I responded to call from a disheartened and confused middle-school teacher in Peterborough, New Hampshire. Her students were not responding to, in fact were rebelling against, a curricular unit on the Northern Forest of New Hampshire that she believed to be relevant and controversial, filled with interesting connections to history, culture, and science.

What was missing? My assessment was that the topic was too disconnected from their everyday lives to feel real. The focus was on northern New Hampshire and not the Monadnock region, where most of her students lived. We found two issues that impacted the local region.

New Hampshire politics are filled with controversy over whether wilderness lands should be preserved or opened for multiuse, which means hunting, logging, and recreation. A local decision was being made to reinstate a statewide hunt to control the growing moose population. Moose were beginning to show up in people's backyards, eating apples, nibbling on ornamental trees, and generally causing a ruckus by wandering onto highways and causing car accidents. The development of a local wetland, necessary for sustaining the integrity of a local wilderness habitat (the Supersanctuary), was also at issue. This confluence of events provided a way to bring the Northern Forest issues home.

We saturated the students with information and current events. Community members shared their expertise and opinions by coming into the school. We traveled locally to visit the highlands and wetland area, to listen to U.S. Fish and Wildlife representatives, and to learn about regional biodiversity issues at a local environmental education center. Dividing the 30 students into six teams, we assigned each team the perspective of a different stakeholder, including hunters, developers, conservation groups, preservation groups, indigenous wildlife (including predators), and the moose population.

A study question that started to emerge from the group focused on the coexistence of wildlife and humans in the Monadnock region. Could we reach consensus on future land use and development while minimizing its impact on wildlife populations? The students suggested a town meeting to debate the issue publicly. I worked with a group of boys interested in the hunter's perspective.

The 10 boys and I went into the woods to better understand the perspective of hunters. We were walking as silently as we could through a hemlock forest that edged a threatened wetland looking for sign of moose and bear. These 10 rowdy "at-risk" 14-year-olds knew how to navigate the thick underbrush and prickly branches, moving stealthily and confidently. The forest seemed to be the milieu that supported and nurtured their learning style, setting them at ease and utilizing their skills.

Most of the boys had fathers or uncles who hunted. They were checking out the territory for game, carefully examining the ground for tracks. In a period of quiet stalking, one of the boys gasped and pointed. There in the shadows stood a huge, antlered form. Calmly stripping the bark from a hemlock, the moose watched our movements. We were caught in the danger zone, and the moose seemed to sense its advantage. I whispered, "Let's back slowly out of this dense cluster of trees and run for it." "How can you leave?" one boy responded. "This is the moment we all wait for. It's awesome. We are caught in his world, at his mercy. The tables are turned because I don't have a gun, and he knows it!"

We squatted for what seemed like hours. I took in the reverence on the faces of these supposedly "out-of-control" boys and marveled at their restraint: they barely moved while intently watching this creature fill its belly. What could their teachers learn from this state of grace? How might this experience shed light on the "hyperactivity" and "restlessness" the boys exhibited in the classroom? Here they were focused, interested, and engaged. They had a reservoir of knowledge and understanding about the natural world. Awe replaced surliness. Joy replaced cynicism. "That was real," said John. "So real I could taste it in the back of my mouth. The chance of a lifetime. I just wanted to suck it all in."

The boys were surveying part of a wildlife corridor called the Supersanctuary. It is a mosaic of wild and scenic ponds and lakes, wetlands and meadows, mountain tops and managed forests in the central highlands of the Monadnock region. You can see its edges from the schoolyard, and most of the students either lived on its edge or had walked on this land. This wetland, known as the Robb Reservoir, was key to the integrity of this 8,000-acre wildlife sanctuary, and it was targeted for a housing development.

The six groups of students addressed the perspectives of their stakeholders and devised a plan for the fate of this "super" wetland and its inhabitants. Students stated their cases in concise, eight-minute presentations at a tightly moderated town meeting. The roles adopted by students seamlessly merged with their personas as they tried out new identities, "stepping into another's shoes." Role playing gave them license to adopt strong values and act them out dramatically without fear of retribution. In the town meeting, the students could let their emotions flow—dissenting and arguing in a safe environment. When the edges got too raw and tender, too personal, or too out of control, adults moderated to get the presenters back on track. The meeting had a serious purpose, as powerful adults were in the audience listening and absorbing their opinions, weighing the results of their research and conclusions.

The developers, first up to speak, described how they planned most of the house lots on half the land, with the other half preserved as forest. The moose, accentuated by a life-size replica in papier-mâché, urged that the animals be left alone and argued that humans would never be satisfied with a small sector if history foretold the future. They singled out the hunters: "You killed us off once, and we're not going to let it happen again!" one girl said. The hunters responded that hunting-license fees helped make the population recovery possible. "We're trying to keep the population balanced," said a student sportsman. "Yeah, well maybe we don't need your help!" yelled the moose-girl.

The preservationists took the developers and hunters to task. They pushed for a return of predatory animals, such as wolves and mountain lions to naturally control the moose population. "The land belonged to the animals first, and when humans tore down their forest, we took away their home, their source of food, and their mating places. We screwed everything up, so how are we qualified to fix anything? We should get the heck out of there." The predator group took the perspective of animals denied access to necessary habitat: "If you think it sounds so good, why don't you humans try being cooped up with no place to run and hunt. Leave us alone!"

Breaking into committees with one representative from each of the interest groups, students took their arguments into a more manageable arena. The fire of this meeting ignited in even typically ambivalent students. Some of the discussions escalated into emotional chaos, full of tension, frustration, and tears, as each student clamored for the right to air his or her opinion.

Mentoring adults struggled to pick out threads of consensus and wend their way toward a palatable resolution. Teachers insisted that students give pause to emotion and take time to reflect before responding. The result was a statement that would be sent to the state legislature, to a New Hampshire Fish and Wildlife Division representative, and to city planners. The caption in a local paper summed it up, "Local students find it's tough to please all in a democracy." This is the statement arrived at by the students:

Resolution
Humans and wild animals must learn to coexist on the same land. Land is limited, and this means we have to control the human use and development of it. One solution is to stop expanding the development of human communities and begin reusing. We must start reusing old lots and abandoned buildings before plowing down the animal's homes by cutting down forest. We think housing developments should be built in phases to see if the homes will sell first before building more. Residents of each housing development near a wilderness must adhere to guidelines that help them respect and get along with wildlife. They have to agree to share the land. We have to let wildlife manage themselves by figuring out how to balance their needs because things are a lot more complicated than humans think they are.

The discussions of human and animal rights were powerful. The result was a consensus that humans must include the needs of other species in their land-use planning and must not presume to have designated control over the lives of animals. The students stated that we must learn to cohabit land with the best interests of all species in mind.

Adolescents and the Design of a Zoo Exhibit

Sneaking around the silent zoo, before "keepers" started their rounds, had adrenaline running high among this group of teenagers. I had been handed the key to a small side gate to bring this group in and out of the grounds. There is something magical about the time before the gates open, some privilege in seeing exhibited animals before they are put on stage.

These 11 teenagers, all 15 years old, were spending four days at the Brookfield Zoo as the culmination of a long interdisciplinary curricular unit. This group worked with me for eight weeks on a special interdisciplinary project that was the result of consulting work I was doing at the Brookfield Zoo in Chicago. I suggested to zoo staff that adolescents might supply them with a critical developmental perspective for the refurbishing of school-based programming and the construction of a new Children's Zoo. Conversations during our early morning walk ranged from awe-inspired observations of animal behavior to angry ethical responses to captivity, the need for zoos in our culture, and the dominance of humans over other species.

One of the research topics that captivated the teens was "How can a zoo cultivate an ethic of care in zoo visitors?" We dug into this subject matter, which was tangled with controversial questions. The outcome was the design of exhibit prototypes that would provoke visitors to take the perspective of the animals. The teens wanted to help people understand the peril of diminishing habitat, the hardships of captivity, and the human impact on other species. Consider what they came up with:

- A glass-bottomed model of the Illinois River, half sparkling clean and half choked with pollution in which participants crossed the river as frogs, ducks, or fish exposed to the benefits of healthy river systems or the hazards of polluted water.
- A three-story rain forest teeming with wildlife that used a treasure-hunt format to guide people through the ecological uniqueness of the canopy, the understory, and the soil. Each student was assigned a predator or prey identity that determined where and how the walk was negotiated.
- A walk-through model of an oak tree, populated with insects and other creepy-crawlies that make their homes in a trunk. At the exit was an exaggerated beehive where participants did the "wiggle dance," searched for pollen, or fed the queen bee.

In their own words, the students had the following goals:

- Be honest about the habits and needs of animals. Talk about predator and prey relationships or show them in a regulated way.
- Show the impact of humans on the environment of animals by showing how human use of resources has gradually diminished the space, food, shelter, and roaming room of animals.
- Have children simulate interactions that would be experienced by animals, seeing the world from the animal's perspective by having to find food sources, smelling out predators, watching for motion in the grass, going down into a den, and playing with siblings.

I took the students to the Brookfield Zoo to explain their exhibits first-hand to zoo staff and designers. The zoo also seemed a perfect venue for exploring deep connections with other biological life forms. We already had grappled with the intellectual issues of captivity and human domination of nature. The intense exposure to animals was stirring empathetic responses: "If I have the same feelings and urges as the baboon, why are they locked up and I'm not?" "I want to hang out in there with them, let loose, and howl!"

A few statements captured early in the project show inklings of a burgeoning awareness that fully bloomed during our visit to the zoo:

> When I was younger, I went to a zoo and watched the penguins being fed. I remember the "keeper" throwing fish and whistling for the penguins to "come and get it." For a long time, I thought that's how penguins ate in the wild. Pretty misleading, don't you think? I think children have an opportunity to experience the lives of animals that they won't normally see, and it should be real, at least as real as it can be. (Annie, age 15)

> When I was little, I went to the Cincinnati zoo all the time. I liked it as a little kid. I could watch the animals play, swim, climb, and be real. But as I grew older, I started to see how the animals were there basically for the public, and it nauseated me. How selfish we were being, locking these creatures up for our entertainment. I would want to make the habitat at the zoo as real as it could be. (Simon, age 14)

The early morning adventure mentioned earlier lasted four hours with no break in the attentiveness of the group. One event in particular captures the essence of our experience. The elephants wandered over to within a foot of the kids, staring at them for the longest time. The female picked up a stick and threw it over the fence. In unison, we all gasped. "Oh, my god, she talked to us!" one of the girls remarked, tears running down her face. "What are you doing in there?" She continued softly speaking back to the elephant: "You probably have babies somewhere, maybe a mate, and you're probably wondering if you'll ever see the jungle again, your family, the place where you feel most at home. What right do we have to do this? I feel so guilty." These genuinely compassionate feelings were difficult to witness.

Always the educator, I followed up later: "How do you think that experience informed your opinion about humans in relation to other species?" One student responded, "I think it showed me just how little difference there is between us. I could feel her spirit like a wave over me when she looked through the fence. It tore me apart to think that humans have claimed some kind of superiority. How are we more deserving of freedom than she is? How can I do anything to preserve her right to walk free in her own place? It is very disturbing."

These are the moments of entry that every educator should look for, unanticipated moments of dissonance and possibility, offering a splendid glimpse at affiliation with nature. Values were being analyzed for relevance and authority. Highly emotional experiences were reviewed and analyzed for deeper meaning. Rich momentary episodes had deep metaphorical repercussions. These unscheduled events are gateways.

Conclusion

Each of the school-based projects described above was poised to take full advantage of the possibility that a student might reach a new threshold of understanding self in affiliation with nature. Adolescents need to partake in controversy and to revel in newly formed perspectives. They have to find safe places to try out their new ideas and make mistakes. They have to have a chance to open themselves to their "wild nature" and howl without scaring anyone. In turn, we, as adults, need to ask adolescents the right questions and be open to their innovative and often unsettling answers. At times, we too should provoke dissonance and recognize that the finest educational opportunities often appear unplanned. We must be ready to receive them, providing a safe and compelling structure in which to explore the dimensions of self and self in relation to nature.

The sparkling moments of awe and wonder, the deep shudders of reverence, the churning beat of sexuality, and the rude spurts of impulse have convinced me that young people in this age group can raise the veil and grasp their "wild" nature with vigor. As an educator, I am searching for ways to use this developmental portal to nurture a more comprehensive human identity, one of self in affiliation with nature. Education with this dynamic in mind will link adolescents inextricably to ecological processes and their biological identity. I believe that it will also increase compassion for other species and caring for the earth.

References

Cobb, E. (1978). *The ecology of the imagination in childhood*. New York: Columbia University Press.

Matthiessen, P. (1978). *The snow leopard*. New York: Viking Press.

Shepard, P. (1996). *Traces of an omnivore*. Washington, DC: Island Press.

Sobel, D. (1993). *Children's special places: Exploring the role of forts, dens and bush houses in middle childhood*. Tucson: Zephyr Press.

Thomashow, M. (1995). *Ecological identity*. Cambridge, MA: MIT Press.

11

Political Economy and the Ecology of Childhood

David W. Orr

We are shocked when violence erupts in schoolyards or when a six-year-old child kills another in cold blood. But the headlines that sensationalize such tragedies reveal only the tip of what appears to be a larger problem that, given our present priorities, will only intensify. Youthful violence is symptomatic of something much bigger evident in diffuse anger, despair, apathy, the erosion of ideals, and the rising level of teen suicide (up threefold since 1960). Nationwide, 17 percent of children are on Ritalin, an antidepressant. Adults often respond with rejection and hostility, making a bad problem worse. We hire more psychologists and sociologists to study our children and more counselors to advise them in things such as "anger management." As a result there are libraries of information about childhood, child psychology, child health, child nutrition, child behavior, and dysfunctional families, much of it quite beside the point. Then in desperation we hire more police to lock up violent youth. We are crossing into a new pattern of relations between the generations, and much depends on how well we understand what is happening, why, and what is to be done about it.

The deeper causes of this situation are not apparent in the daily headlines and news reports. Dysfunctional families, depression, youthful violence, and the rising use of chemicals to sedate children are symptoms of something larger. Without anyone saying as much and without anyone intending to do so, we have unwittingly begun to undermine the prospects of our children, and I believe that at some level they know it. This essay is a meditation on the larger patterns of our time and their effects on children. My argument is that the normal difficulties of growing up are compounded, directly and indirectly, by the reigning set of assumptions, philosophies, ideologies, and even mythologies by which we organize our affairs and conduct the business of society—what was once called *political economy*. The study of political economy began with Adam Smith and continued on through Karl Marx to the present with scholars such as Yale University political scientist Charles Lindblom. Due to academic specialization and diminished public involvement in politics and community life, the field has declined. As a result, we have increasing difficulty in discerning larger social, economic, and political causes of our problems and doing something constructive about them. This essay is an attempt, in effect, to connect the dots describing those larger patterns. The first section below reviews evidence about the intersection of childhood and political economy from many different perspectives. The second section is a more explicit rendering of the political economy of contemporary global capitalism. The final section sketches some of the alternative political and economic arrangements necessary to honor our children and protect future generations.

The Evidence

Health

By one estimate average young Americans carry at least 190 chlorinated organic chemicals in their fatty tissues and blood and another 700 additional contaminants as yet uncharacterized. Nursing infants in their first year of life have a higher body burden of dioxin than the average 70-year-old man (Thornton, 2000, pp. 41–43). They are threatened by the air they breath, the food they eat, the water they drink, many of the materials common to everyday use, and fabrics in the designer clothes they wear. We have subjected our children to a vast experiment in which their body chemistry is subjected to hundreds of chemicals for which we have no evolutionary experience. We have good reason to suspect that their ability to procreate is being threatened by dozens of commonly used chemicals that disrupt the normal working of the endocrine system. As a result sperm counts are falling, and incidences of reproductive disorders of various kinds are rising (Colborn, Dumanoski & Myers, 1996). We have reason to believe that exposure to some kinds of chemicals can cause varying levels of damage to the brain and the nervous system. We have, in short, every reason to believe that a century of promiscuous industrial chemistry is seriously effecting our children. And we have reason to believe that current trends, unless altered, will grow worse. The scientific evidence is compelling but is widely dismissed because of a kind of deep-seated denial and a mindset that demands absolute proof of harm before remedial action can be taken. So instead of eliminating the problem, we quibble about the rate at which we can legally poison each other.

Much of the same can be said about exposure to heavy metals. Nearly a million children under the age of five still suffer from low-level lead poisoning (*Environment and Health Weekly*, 2000, no. 687). Half of all children in the United States have lead levels that impair reading abilities ("Living on Earth," 2000). Even though leaded gasoline has been phased out, Americans still have "average body burdens of lead approximately 300 to 500 times those found in our prehistoric ancestors" (*Environment and Health Weekly*, 2000, no. 689). The problem is not that we do not know the effects of lead and other substances on the human mind and body but that corporations have the power to control public policy long after evidence of harm is established beyond reasonable doubt (Kitman, 2000).

Food and Health

More children exhibit the effects of bad diet and lack of exercise than ever before. The average diet of children has deteriorated in this age of affluence and fast food. Of those under the age of 19, one-quarter are overweight or obese. The U.S. Surgeon General believes that the problem is epidemic: "we see a nation of young people seriously at risk of starting out obese and dooming themselves to the difficult task of overcoming a tough illness" (Critser, 2000, p. 150). Children are bombarded with 10,000 advertisements each year hawking fatty and sugar-laden food. The problem with a junk-food diet is not just obesity but the long-term damage it does to the pancreas, kidneys, eyes, nerves, and heart. There is a national eating disorder fostered by the corporations that feed us. But the disorder is not evenly visited on children. It is most apparent among children from lower-class homes. The junk diet of fat-laden fast foods represents a kind of class warfare in which corporations prey on the gullible, the poor, and the defenseless.

The problem of diet is compounded by a decline in physical exercise. One expert estimates that amount of physical activity of the typical child has declined 75 percent since 1900 (Healy, 1990, p. 171). Another study shows a sharp decline in the average time children between the ages of three to 12 spend out of doors from an average of 86 minutes per day in 1981 to 42 minutes in 1997 (Fishman, 1999, appendix). Indeed, capitalism works best when children stay indoors in malls and in front of televisions or computer screens. It loses its access to the minds of the young when they discover pleasures that cannot be bought.

Information

The average young person watches television a little over four hours per day. They are exposed daily to the most tawdry kinds of "entertainment" and advertisements. Corporations spend $2 billion each year targeted specifically on the young, intending to lure them into a life of unthinking consumption. The American Academy of Pediatrics estimates that by age 18 they have seen 360,000 television advertisements and 200,000 violent acts (*Environment and Health Weekly*, 2000, no. 681). We have no good way to estimate the cumulative impact of these images on the growing human mind, but we may reasonably surmise that television strongly effects what they know, what they pay attention to, and what they can know and pay attention to. We have, by one estimate, over 1,000 studies showing that "significant exposure to media violence increases the risk of aggressive behavior in certain children and adolescents, desensitizes them to violence and makes them believe that the world is a 'meaner and scarier place' than it is" (ibid.). We know, too, that young people on average can recognize over 1,000 corporate logos but only a handful of plants and animals native to their places. They are probably less adept with language than previous generations. They are increasingly hooked on the Internet so that some colleges have had to hire counselors to deal with the problem as an addiction. And what has not happened in all the TV and Internet watching? The list is a long one—healthy contact with adults, making friends, outdoor exercise, reading, contemplation, and creative activity.

Education

With growing numbers of dysfunctional families, schools are now expected to make up for what parents ought to do. At the same time, schools and colleges are under increasing financial pressures and have increasingly become places of commerce. Many children are now exposed to the blatant commercialization of Channel One during school time. Many are required to read text materials developed by corporations that celebrate the virtues of capitalism without acknowledgment of its vices. More and more they are educated to take proficiency tests, not to learn creatively and critically. While we talk about the importance of learning, public spending tells a different story. A city like Cleveland, with one of the worst urban school systems in the nation, can find hundreds of millions of dollars for a new football stadium used eight times a year but not the money or the foresight to repair the leaking roofs of its public schools. Nationally, some 60 percent of our schools need repair (Healy, 1998, p. 92). Young people are quick to comprehend adult priorities. Financial priorities in higher education are also skewed. Commerce is making deep inroads into the academy, and colleges and universities have become heavily dependent on corporate support. As a result, corporations have acquired unprecedented influence over whole departments and the evolution of entire disciplines (Press & Washburn, 2000).

Technology

A rising percentage of young people now spend many hours each day on the Internet or playing video games. Signs of trouble are already apparent. Internet addiction is a serious and growing problem. One study has shown that even a few hours a week online caused a "deterioration of social and psychological life" and higher levels of depression and loneliness among otherwise normal people (Harman, 1998). These symptoms are also by-products of the mental disorientation caused by overexposure to a contrived electronic reality. As the technology for simulation advances, we may expect that the young so exposed will find it increasingly difficult to distinguish the contrived from the real, to establish deep emotional ties to anyone or anything, or simply to take responsibility for their own actions.

In the not-too-distant future, researchers in artificial intelligence and robotics are planning to create self-replicating machines that will be more intelligent than humans. Evolution, they say, works by replacement of the inferior by the superior and unabashedly regard themselves as the agents of evolution with a mandate to create the next stage of intelligent life. It is not at all far-fetched to think that such alien intelligence could well find humans—meaning our children and grandchildren—inconvenient (Joy, 2000). This is no longer some distant science fiction but reality coming inexorably into view.

If the present technological developments lead to a world of simulated reality that is more real to some in the next generation than actual experiences, then it is also increasingly possible that advances in fields such as artificial intelligence will diminish what it means to be human.

Ecology and Climate

The numbers are staggering. In the United States alone, we lose more than a million acres each year to urban sprawl, parking lots, and roads. We continue to destroy tropical forests worldwide at a rate of 80,000 square miles per year (Leakey & Lewin, 1995, p. 237). The rate that we are driving species extinct rivals that of the last great extinction spasm 65 million years ago. Oceans and virtually every ecosystem on the planet are now deteriorating due to human activity. The scientific evidence indicates that climatic change is happening more rapidly than thought possible even a few years ago. Biotic impoverishment, climatic change, and pollution are beginning to undo millions of years of evolution and with it the rightful heritage of our children.

Despite a burgeoning global economy, the plight of children worldwide is much worse than it was a generation ago. In some cities it is now common to see street children with no known parents and no home other than the street. They are sometimes killed or persecuted by police and preyed on by those who exploit them shamelessly. It is common for children in developing countries to work under sweatshop conditions making products for global corporations. In Africa, the Balkans, the

Middle East, and Ireland, the facts differ from place to place but only as variations on a common theme of abuse, neglect, exploitation, and an astonishing level of intergenerational incompetence.

It is ironic that adults do not like the children they are raising. By one accounting only 37 percent of adults believe that today's youth will "make this country a better place." Two-thirds of the adults surveyed find young people rude, spoiled, violent, and irresponsible (Applebome, 1997). Ninety percent believe that values are not being transmitted to the young. And only one in five believed it common to find parents who are good role models for their children. Previous generations often regarded the young with skepticism, but what is different now, according to the authors of this study, is the intensity of antagonism between the generations and the empirical evidence supporting it. Daniel Goleman (1985), author of *Emotional Intelligence*, estimates that American children have declined on some 40 indicators of emotional and social well-being. (cited in Healy, 1998, p. 174).

Perhaps I have exaggerated the problems, and the prospects for our children are quite different than I have described. Maybe these problems are mostly unrelated and arise from different causes. As any reader of Charles Dickens knows, children in earlier times were sometimes badly treated and lived in harsh conditions. And children from affluent homes are certainly not exposed to many hardships characteristic of some earlier times. But the evidence, in its entirety, is so well documented and so pervasive that we cannot mistake the larger pattern without thoroughgoing self-deception. We are unwittingly undermining the health of ecosystems, a sense of commonwealth, hope for a decent future, and our children's physical health, mental health, connection to adults, sense of continuity with the past, and connections to nature. But we have difficulty in seeing whole systems in a culture shaped so thoroughly by finance capital and narrow specialization. However bad the situation of children in the past, no generation ever has done, or could have done, such systematic violence to its progeny and their long-term prospects. Most would adamantly protest that they love their children and are working as hard as possible to make a good life for them, and I believe that most parents and adults fervently believe that they are doing so. But we are caught in a pattern of deep denial that begins by confusing genuine progress, a difficult thing to appraise, with what is simply easy to measure—economic growth. We confuse convenience and comfort with well-being, longevity with health, Scholastic Aptitude Test scores with real intelligence, and a rising gross national product with real wealth. We express our affection incompetently. Without anyone intending to do so, we have launched a raid on our children's future stealing things not rightfully ours, leaving behind a legacy of destruction and degradation—a kind of intergenerational scorched-earth policy. But why?

Political Economy

The conditions in which children experience nature are in large part an artifact of political economy, which Michael M'Gonigle defines as "the study of society's way of organizing both economic production and political processes that affect it and are affected by it" (1999b, p. 125). Beginning with Adam Smith and later Karl Marx, the study of political economy has aimed "to uncover and explain what might be called the 'system dynamics' of a society's processes of economic and political self-maintenance" (ibid., p. 126). The political economy of the modern world, in this view, is organized around the pursuit of economic growth, a science presumed to be value neutral, and the institutions of the state and corporation. Its ideology is "high modernist," which in political scientist Jack Scott's (1998, p. 4) words means

> a muscle-bound, version of the self-confidence about scientific and technical progress, the expansion of production, the growing satisfaction of human needs, the mastery of nature (including human nature), and, above all, the rational design of social order commensurate with the scientific understanding of natural laws.

The main features of modern political economy are well known even if their effects on childhood are not. The first and most obvious feature of contemporary political economy is the belief in the importance of economic growth and material accumulation. One day the major political fault line in the twentieth century about whether growth was to be organized by markets or governments will be seen as a minor doctrinal quibble. Regardless of specifics, economic growth has become the central goal for virtually every national government. Election outcomes are now more than ever an artifact of short-term economic performance. A second feature of modern political economy is the centrality of the global corporation. We are now provisioned with food, energy, materials, entertainment, health, livelihood, information, shelter, and transport by global corporations that operate with little oversight. The economic scale of the largest corporations dwarfs all but the largest national economies. As a result, corporations dominate national politics and policy and, through relentless advertising, the modern worldview as well. A third component of contemporary political economy is a particular kind of science rooted in the thinking of Descartes, Galileo, Bacon, and Newton. That science presumes a separation of subject from object, humankind from nature, and fact from value. Its power is derived from its ability to reduce the objects of inquiry to their component parts. Its great weakness has been its inability to associate the knowledge so gained into its larger ecological, social, cultural, and normative contexts.

Political economy organized on these three pillars has many collateral effects on children. First, a society organized around economic growth is one that is in constant turmoil. Austrian economist Joseph Schumpeter described the process by which physical capital is rendered obsolete as "creative destruction." Economic growth, then, means that the old and familiar are continually being replaced with something new and more profitable to the owners of capital. Similarly, the growth economy and the continual battle for market share among corporations is driven by and in turn drives a process of incessant technological change aiming for greater efficiency and speed. Creative destruction and technological dynamism, in turn, increase the velocity of lived experience. Not only is rapid change regarded as good, but rapid movement is as well. Corporations not only sell things: they sell sensation, movement, and speed, and these, too, are integral to the growth economy.

Little attention has been given to the effects of creative destruction, technological change, and increased velocity on the development of children, but it cannot be insignificant. For one thing, familiar surroundings and places where the child's psyche is formed are subject to continual modification, which is called *development* but which to the child is a kind of obliteration. These places, regarded as real estate to the capitalist mind, are where the child forms its initial impressions of the world. They are, as Paul Shepard (1996) noted, the substrate for the adult mind. Some part of otherwise inexplicable teenage behavior in recent decades may be a kind of submerged grieving over the loss of familiar places rendered into housing tracts or shopping malls (Windle, 1994). The effects of technological change and the consequent increase in the speed of lived experience on children is largely unknown, but it is reasonable to think that the healthy pace of human maturation is much slower than the frenetic speed of a technological society. The problem of speed is, I think, pervasive. At one level exposure to television (averaging more than four hours per person per day) with constantly changing images effects the neural organization of the mind in ways we do not understand. At another level, the decline in time spent with children means that parenting is compressed into smaller and smaller chunks of time. In either case, the child's sense of time is bent to fit technological and economic imperatives.

A second collateral effect arises from rampant materialism inherent in the growth economy. Childhood lived in more austere times was no doubt experienced differently from one lived in seemingly endless abundance. From birth on, children in an affluent culture marinate in a surfeit of things as well as in the desire for things not yet possessed. Love in the growth economy is increasingly expressed by giving things and not by spending time with the child. Again, we have little idea of the long-term effects of excessive materialism on the child, but it is reasonable to think that its hallmarks are satiation, shallowness, and the loss of deeper feelings having to do with a secure and stable identity rooted in the self, relationships, and place. The important fact is not simply the effects of materialism but the more complex effects of the worldview conveyed in relentless advertising that hawks the message of instant gratification in a world of endless abundance. Whatever its other effects on the child, nature in a culture so lived can only recede in importance. Time once spent doing farm chores, exploring nearby places, fishing, or simply playing in a vacant lot has been replaced by the desire to possess or to experience some bought thing. It is, again, not far-fetched to think that one consequence of materialism is a loosening of ancient ties to place and an acquaintance with wildness. Nor is it unreasonable to suppose that the effect of several decades of glorifying money and things is now apparent in polls showing that the young increasingly want to get rich rather than live a life of deeper purpose.

A third collateral effect of contemporary political economy is that the world is increasingly rendered into commodities to be sold. Indeed, this is the purpose of the growth economy. Having saturated the market for automobiles and washing machines, it proceeded to sell us televisions and stereo equipment. Having saturated those markets, the economy moved on to sell us computers and cell phones. Eventually, it will sell us a version of reality that aims to supplant more than most of us care to admit. Commodification, too, has its effects on the ecology of childhood. Those things that people once did for themselves as competent citizens or as self-reliant communities are now conveniently purchased. What's good for the gross national product, however, is often detrimental to communities. Real community can be formed only around mutual need, cooperation, sharing, and the daily exercise of practical competence. The effect of the growth economy and corporate dominance is to undermine the practical basis for community and with it the lineaments of trust. The absence of these qualities cannot be seen and so cannot be easily measured. Nonetheless, by many accounts there is a marked decline in community strength and social trust that cannot leave childhood unaffected (Putnam, 2000). I suspect that these are mostly manifest in a decline in the imagination of a world of rich social possibilities that are lived out in real communities by people who have learned to live in interaction and not in isolation. Instead, the young are socialized into an increasingly atomized world of extreme individualism governed by the assertion of freedoms without rights. As such they are being trained to become reliable, even exuberant, consumers but inept citizens and community members.

Much of the same can be said about the effects of economic growth on child care and the evolution of emotionally grounded intelligence in children. Economic necessity often forces both parents to work, leaving less time with their children. In psychiatrist Stanley Greenspan's (1997, p. 179) words, one result of these social adaptations to economic forces is that "our nation has . . . launched a vast social experiment . . . and the early data are not encouraging." What's at risk, he believes, are the "relationships on which developmental patterns rest" in a society in which "intimate personal interaction is declining and impersonality is increasing" (ibid., 169). These relationships, however, are crucial for the development of emotionally grounded intelligence.

Fourth, contemporary political economy is rooted in the tacit acceptance of high levels of risk that both jeopardize the lives of children and color their worldview. The growth economy creates mountains of waste, much of it toxic and some of it radioactive. It has been the driving force behind biotic impoverishment and the loss of biological diversity. Its further expansion now threatens climatic stability. Risks from technology and the scale of the economy are now pervasive, global, and permanent (Beck, 1992). But the response of mainstream science, reflected in the practices of cost-benefit analysis or risk analysis, is rooted in the same kind of thinking that created the problems in the first place (O'Brien, 2000). We have no way to know the full range of biophysical effects on children, nor can we say with certainty how they perceive the tapestry of risk that shrouds their future. But again, it is reasonable to think that these risks contribute to an undertone of despair and hopelessness.

Finally, the role of science in this larger political economy resembles more and more what Wendell Berry (2000, p. 18) calls "modern superstition," in which "legitimate faith in scientific methodology seems to veer off into a kind of religious faith in the power of science to know all things and solve all problems." Increasingly, children grow up in a thoroughly secular culture, often without awareness that life is both gift and mystery. They are, in other words, spiritually impoverished. Since humans cannot live without meaning, their search for meaning, bereft of the possibility for authentic expression, can take ever more bizarre and futile forms.

The situation of some children has improved vastly since the early years of capitalism, when child labor was common. Improved living circumstances for some children fortunate enough to be raised in middle- or upper-class homes is a reality, with all of the caveats noted above. But little in contemporary political economy mandates that incomes will be fairly distributed or that children in other cultures will not be exploited to produce cheap sneakers and designer jeans for those living in affluence. Nor does this political economy afford any child living in the future adequate protection from pollution, reproductive disorders, overexploitation of resources, climatic change, or loss of species.

Relative to their relation to nature, the reigning political economy has shifted the lives and prospects of children from

- Direct contact with nature to an increasingly abstract and symbolic nature,
- Routine and daily contact with animals to contact with things,
- Immersion in community to isolated individualism,
- Less violence to more violence, much of it vicarious,
- Direct exposure to reality to abstraction and virtual reality, and
- A relatively slow pace to a fast pace.

There are certainly exceptions—the Amish, for example, who are notable because they are exceptions. On balance, children in modern society are heavily shaped by a contemporary political economy that stresses materialism, economic growth, and human domination of nature and that tolerates large-scale ecological risks with irreversible consequences. Children's view of nature is increasingly distant, abstract, and utilitarian. However affluent, their lives are impoverished by diminishing contact with nature. Their imaginations, simulated by television and computers, are being impoverished ecologically, socially, and spiritually. The young, in Neil Postman's (2000, pp. 125–126) words, have been rendered into an "economic category . . . an economic creature, whose sense of worth is to be founded entirely on his or her capacity to secure material benefits, and whose purpose is to fuel a market economy." This is not happening according to any plan; it is, rather, the logical outcome of the regnant system of political economy.

We have, in other words, created a global system of political economy in which it is not possible to be faithful or effective stewards of our children's future. It is a system that, by its nature, clogs many of its children's arteries with fast food. It is a system that, by its nature, poisons all of its children, albeit unevenly, with chemicals and heavy metals. It is a system that, by its nature, must saturate most of their minds with television advertisements and electronic trash. It is a system that, by its nature, must impoverish ecosystems and change climate. It is a system, that by its nature, undermines communities and family ties. It is a system, run by people who love their children, that will measure risks to them with great precision but is incapable, as it is, of implementing alternatives to risky things. It is a system that, by its nature, must remove most children from direct contact with unmanaged nature. And it is a system that encourages people to see the problems that arise from its very nature as anomalies—not as parts of a larger and deeply embedded pattern. We have unwittingly created a global political economy that prizes economic growth and accumulation of things above the well-being of children.

The important issues for our children are not narrowly scientific. They have little to do with symptoms and everything to do with systems. What kind of changes in the system of political economy would be necessary to protect the rights and dignity of children now and in the future?

A Child-Centered World

On July 30, 1998, the Supreme Court of the Philippines in *Minors Oposa* ruled that a group of 44 children had standing to sue on behalf of subsequent generations. In their suit, the children were trying to cancel agreements between timber companies and the Philippines government. The Court found "no difficulty in ruling that they can, for themselves, for others of their generation, and for the succeeding generations, file a class suit . . . based on the concept of intergenerational responsibility insofar as the right to a balanced and healthful ecology is concerned" (Gates, 2000, p. 289; Ledewitz, 1998, pp. 605–606). The court considered the essence of that right to be the preservation of "the rhythm and harmony of nature," including "the judicious disposition, utilization, management, renewal and conservation of the country's forest, mineral, land, waters, fisheries, wildlife, off-shore areas and other natural resources." The Court further stated that "every generation has a responsibility to the next to preserve that rhythm and harmony for the full enjoyment of a balanced and healthful ecology." That right, the Court argued, "belongs to a category . . . which may even predate all governments and constitutions . . . exist[ing] from the inception of humankind." Without the protection of such rights "those to come [will] inherit nothing but parched earth incapable of sustaining life."

The Court's decision recognizes what is, I think, simply obvious—that the right to a "balanced and healthful ecology" is the *sine qua non* for all other rights. The Court acknowledged, in other words, that human health and well-being are inseparable from that of the larger systems on which we are utterly dependent. The Court's decision implicitly acknowledges the inverse principle that no generation has a right to disrupt the biogeochemical conditions of the earth or to impair the stability, integrity, and beauty of biotic systems, the consequences of which would fall on subsequent generations as a form of irrevocable intergenerational "remote tyranny."

No mention of ecological rights was made in our own Bill of Rights and subsequent constitutional development because, until recently, only the most prescient realized that we could damage the earth enough to threaten all life and all rights. But the idea that rights extend across generations was part of the revolutionary ethos of the late eighteenth century. The Virginia Bill of Rights (June 12, 1776), for example, held that "all men . . . have certain inherent rights, of which when they enter into a state of society, they cannot by any compact deprive or divest their posterity; namely, the enjoyment of life and liberty, with the means of acquiring and possessing property, and pursuing and obtaining happiness and safety" (Commager, 1963, p. 103). That same idea was central to Thomas Jefferson's political philosophy. In the famous exchange of letters with James Madison in 1789, Jefferson (1975, p. 445) argued that "the earth belongs in usufruct to the living . . . no man can, by natural right, oblige the lands he occupied, or the persons who succeed him in that occupation, to the paiment of debts contracted by him. For if he could, he might, during his own life, eat up the usufruct of the lands for several generations to come, and then the lands would belong to the dead, and not to the living." Jefferson's use of the word *usufruct* (the legal right of using and enjoying the fruits or profits of something belonging to another) is central to his point. For Jefferson, "the essence of the relationship between humans and the earth," in Richard Matthews (1995, p. 256) words, is "that of a trust, a guardianship, where the future takes priority over the present or past." Initially skeptical, Madison, in time, came to hold a similar view (ibid., p. 260). On the other side of the political spectrum, the founder of modern conservatism, Edmund Burke, arrived at a similar position. In his *Reflections on the Revolution in France* (1790), Burke (1986, p. 119) described the intergenerational obligation to pass on liberties "as an entailed inheritance derived to us from our forefathers, and to be transmitted to our posterity." For Burke, society is "a partnership not only between those who are living, but between those who are living, those who are dead, and those who are to be born" (ibid., p. 195).

It is reasonable, given what we now know, to enlarge the concept of intergenerational debt to include intergenerational ecological debts, such as biotic impoverishment, soil loss, ugly and toxic landscapes, and unstable climate. It is entirely logical to believe that the right to life and liberty presumes that the bearers of those rights also have prior rights to the biological and ecological conditions on which life and liberty depend. If Jefferson were alive now, he would, I think, agree wholeheartedly with that amendment. Similarly, Burke would agree that the entailed inheritance of institutions, laws, and customs must also be expanded to include its ecological foundations, without which there can be no useable inheritance at all. This suggests a convergence of left and right around the idea that the legitimate interests of our children and future generations sets boundaries to present behavior and changes the character of the present generation from property holders with absolute ecological rights to trustees for those yet to be born. The echo of this tradition is sounded in our time in documents such as the Brundtland Commission Report (World Commission, 1987, p. 40), *Our Common Future*, which defines sustainable development as a way "to meet the needs and aspirations of the present without compromising the ability to meet those of the future." Similarly, the "Earth Charter" (2000) aims, in part, to "transmit to future generations values, tradition, and institutions that support the long-term flourishing of Earth's human and ecological communities."

Extending rights to some limits the freedom of others, thereby acknowledging that we live in a community and must be disciplined by the legitimate interests of every member of that community, now and in the future. Mesmerized by the industrial version of progress, we have been slow to recognize the revolutionary implications of this idea. But taken seriously, what does children's standing to sue on behalf of the unborn require of us, and how does the extension of certain ecological rights across time affect the choices we make today? The answer is that we are required to follow the thread of obligations back to the economic and political conditions affecting children now and will do so in the future. This requires, in short, that we rethink political economy from the perspective of those who cannot speak on their own behalf.

The most obvious of the present conditions affecting children now has to do with the distribution of wealth. It is an article of faith in the contemporary political economy that everyone has the right to amass as much wealth as he or she possibly can and that any single generation has the same right regarding subsequent generations. As a result the top 1 percent of people living in the United States have greater financial net worth than the remaining 95 percent (Gates, 2000, p. 79). Working-class families watched their real income decline by 7 percent between 1973 and 1998, putting more pressure on children who receive, as Jeff Gates puts it, "less parenting from substantially more stressed parents" (ibid., p. 47). Despite the huge increase in wealth in the past half century, one-fifth of American children still live in poverty (ibid., p. 69). To guarantee that every child has the basics of food, shelter, medical care, decent parenting, and education means that we must address basic problems of economic security for families. Since poverty and its effects are often self-perpetuating across generations, inequity casts a long shadow over the future.

Similarly, implicit in the political economy of capitalism is the faith that the prosperity of the present generation will flow into the future as a positive stream of wealth. Losses in natural capital, it is assumed, will be offset by increased wealth. It is clear, however, that a stream of liabilities—toxic waste dumps, depleted landscapes, biotic impoverishment, climate change—cannot be nullified because natural and economic capital are not always interchangeable (Daly and Costanza, 1992). The intergenerational balance of the economic capital created minus the natural capital lost may not be positive because the costs of repairing, restoring, or simply adjusting to a world of depleted natural capital will exceed the benefits of advanced technology, sprawling cities, and larger stock portfolios.

Second, the recognition of children's rights would require us to rethink the taboo subject of property ownership. From that perspective we are obliged to protect not only the big components of the biosphere but also the small places in which children's lives are lived. Children need access to safe places, parks, and wild areas. This recognition would cause us more often to rebuild decaying urban areas, restore degraded places, preserve more open spaces and river corridors, build more parks, set limits to urban sprawl, and repair ruined industrial landscapes. But doing so would require changing our belief in the nearly absolute rights of the landowner, supposedly derived from English philosopher, John Locke. We need to reread John Locke with the interests of children and future generations in mind. In fact, Locke's case for private ownership carried the caveat that land ownership should be limited so that "there is enough and as good left in common for others" (Locke, 1688/1965, p. 329; Shrader-Frechette, 1993). The rights of children and future generations runs counter to notions of property that give present owners the rights to do with land much as they please. At its most egregious, absentee corporations own land and subsurface mineral rights to large portions of Appalachia while paying minuscule taxes and practicing a kind of mining that decapitates entire mountains (Lockard, 1998, p. 171). Nothing in the law or current business ethics or mainstream economics would require them to give the slightest heed to the rights of the children living in those places or to those who will live there. Property rights in a child-centered political economy will require that owners must leave "enough and as good" or forfeit ownership.

Third, what do the rights of children mean for the interpretation of other rights, such as the First Amendment guarantees of freedom of speech and freedom of the press? From a child's point of view, those freedoms have been corrupted to allow corporations to target children through advertising, movies, and television programming. More fundamentally, they have been corrupted to protect the rights of property, not the rights of people, by allowing corporations the same legal standing as persons. A child-centered political economy would, I think, permit no such reading of the Constitution or violations of common sense. Freedom of speech was intended by the Founders not as a license but as a fundamental protection of religious and political freedoms and should not be interpreted as a right to prey on children for any purpose whatsoever.

Perhaps most difficult of all is the question of what the rights of children mean for the development of technology. Neil Postman (1982, p. 145) once asked whether "a culture [could] preserve humane values and create new ones by allowing modern technology the fullest possible authority to control its destiny." We have good reason to believe that the answer is no. But the subject is virtually taboo in the United States. Biologist Robert Sinsheimer (1978, p. 33) once proposed to limit the rights of scientists where their freedom to investigate was "incompatible with the maintenance of other freedoms." His argument was met with a thundering silence. In a society much enamored of invention, he inconveniently asked whether the rights of the inventor to create risky and dangerous technologies exceeded the rights of society to a safe and humane environment. Nearly a quarter of a century later, computer software engineer Bill Joy raised the same question regarding the rapid advance in technologies with self-replicating potential like genetic engineering, nanotechnologies, and robotics. In Joy's (2000, p. 256) words, "We are being propelled into this new century with no plan, no control, no brakes." Like Sinsheimer, Joy proposed placing limits on the freedom to innovate assuming that the rights of some to pursue wealth, fame, or simply their curiosity should not trump the rights of future generations to a decent and humane world. A child-centered political economy would begin with the rights of the child and future generations and not with those of the scientist and the inventor. It would put brakes on the rights of technological change and scientific research where those might incur large and irreversible risks.

Fifth, a child-centered political economy would give priority to democratically controlled communities over the rights of finance capital and corporations—another taboo subject. In a series of decisions beginning with the 1819 *Dartmouth College* case and culminating in the 1886 *Santa Clara* case, the U.S. Supreme Court has given corporations the same protections given to individuals. In the words of Kalle Lasn and Tom Liacas (2000, p. 41),

We live in the shadow of a super-species, a quasi-legal organism that competes with humans and other life-forms in order to grow and thrive. . . . It can "live" in many places simultaneously. It can change its body at will—shed an arm or a leg or even a head without harm. It can morph into a variety of new forms to absorb other members of its species or be absorbed itself. Most astoundingly, it can live forever. To remain alive, it only needs to meet one condition: its income must exceed its expenditures over the long run.

Corporations now rival or exceed the power and influence of nation-states. The largest 100 control 33 percent of the world's assets but employ only 1 percent of the world's labor (ibid.). They presently control trade, communications, agriculture, food processing, genetic materials, entertainment, housing, health care, transportation, and, not least, the political process. If anything is left out of their control, it is because it is not profitable. Some routinely lie, steal, corrupt, and violate environmental laws with near impunity. As a consequence, there is no safe future for children, nor are there safe communities in a world dominated by organizations that exist partly beyond the reach of law and owing no loyalty to anyone or to any place. The solutions are obvious. Corporations are chartered by the state and can be dissolved by the state for just cause. We have implemented a "three strikes and you are out" standard for criminals. Why not hold corporations and the people who serve them to the same standard? Wayne township in Pennsylvania, for example, bars any corporation with three or more regulatory violations within seven years. Many are asking for community control of investment capital and major assets. Nine midwestern states forbid corporate farm ownership. What attorney Michael Shuman (1998) calls "going local" requires a rejuvenation of democracy beginning by establishing local control over resources and investment decisions.

Finally, as farsighted and revolutionary as the decision of the Philippine Court is, there is another and collateral right to be preserved, which is children's capacity to affiliate with nature and the places in which they live. Biologist Hugh Iltis (quoted in Shepard, 1998, p. 136) describes that capacity in these words:

Our eyes and ears, noses, brains, and bodies have all been shaped by nature. Would it not then be incredible, indeed, if savannas and forest groves, flowers and animals, the multiplicity of environmental components to which our bodies were originally shaped, were not, at the very least, still important to us?

Harvard biologist E. O. Wilson (1984, p. 81) calls this "biophilia," which he defines as "the affinity for life and lifelike processes," and he writes that "we are a biological species and will find little ultimate meaning apart from the remainder of life." Rachel Carson (1956/1984, p. 45) defined it simply as "the sense of wonder" aided and abetted by "the companionship of at least one adult."

Is the opportunity to develop biophilia and a sense of wonder important? Can it be considered a right? To the first question the answer is yes because it is unlikely that we will want to preserve nature only for utilitarian reasons. We are likely to save, as Stephen Jay Gould notes, only what we have first come to love. Without that affection, in other words, we are unlikely to care about the destruction of forests, the decline of biological diversity, or the destabilization of climate. To the second question the answer must again be affirmative because affiliation with nature, by whatever name, is an essential part of what makes us human. We have good reason to believe that human intelligence evolved in direct contact with animals, landscapes, wetlands, deserts, forests, night skies, seas, and rivers. We have reason also to believe, as Paul Shepard (1998, p. 127) put it, that "the potential for becoming as fully intelligent and mature as possible can be hindered and even mutilated by circumstances in which human congestion and ecological destitution limit the scope of experience." We can all agree that the act of deliberately crippling a child would violate basic rights. By the same token, mutilation of a child's capacity to form what theologian Thomas Berry (1999, pp. 15, 82) calls "an intimate presence within a meaningful universe," a thing harder to discern, is no less appalling because it would deprive the child of a vital dimension of experience:

> We initiate our children into an economic order based on exploitation of the natural life systems of the planet. To achieve this attitude we must first make our children unfeeling in their relation with the natural world. . . . For children to live only in contact with concrete and steel and wires and wheels and machines and computers and plastics, to seldom experience any primordial reality or even to see the stars at night, is a soul deprivation that diminishes the deepest of their human experiences.

The result of that deprivation is a kind of emotional and spiritual blindness to the larger context in which we live, abridging the sense of life.

Were we to take the right to a "balanced and healthful ecology" seriously, we would do all in our power to protect the right of children to develop a healthy kinship with the earth. We would honor the ancient tug of the Pleistocene in our genes by preserving opportunities for children to "soak in a place and [for] the adolescent and adult . . . to return to that place to ponder the visible substrate of his or her own personality" (Shepard, 1996, p. 106). We would "find ways to let children roam beyond the pavement, to gain access to vegetation and earth that allows them to tunnel, climb, or even fall" (Nabhan & Trimble, 1994, p. 9). We would preserve the right to "the playful exploration of habitat . . . as well as the gradual accumulation of an oral tradition about the land [that] have been essential to child development for over a million years" (ibid., 83). We would preserve wildness even in urban settings. This is not nature education as commonly understood. It is, rather, a larger subject of how and how carefully we manage the ecology of particular places to permit the full flowering of human potentials.

Conclusion

The invention of childhood in the late Middle Ages was a discovery, of sorts, that the child was not simply a miniature adult but a distinct stage of life with its own needs and developmental pattern (Aries, 1962). More than a useful discovery, it was, rather, a fundamental acknowledgment that a decent culture needed to make a greater effort to shelter, nourish, and establish individual personhood than had previously been the case. We have good evidence from many sources that childhood as a distinct and protected phase of life is disappearing, and we have every reason to fear that loss. The primary cause is an errant system of political economy loosed on the world. It is failing children now and will in time fail catastrophically. Children will bear the brunt of that failure as well. Far from having settled all of the big political and economic issues, we have yet to create a political economy that protects the biosphere and the physical, mental, emotional, and spiritual well-being of children and through them the future of our species. Hopefully, we are at the beginning of what Thomas Berry (2000, p. 55) calls the Ecozoic Era, "when humans will be present to the Earth in a mutually enhancing manner."

For that hope to become manifest we must first organize our political and economic affairs in a way that honors the rights of all children. The irony of our situation is that what appears from our present vantage point to be altruism will, in time, come to be seen as merely practical, far-sighted, self-interest.

References

Applebome, P. (1997, June 26). Children score low in adults' esteem. *The New York Times*, p. A25.

Aries, P. (1962). *Centuries of childhood*. New York: Vintage.

Beck, U. (1992). *Risk society*. Beverly Hills: Sage.

Berry, T. (1999). *The great work*. New York: Bell Tower.

Berry, W. (2000). *Life is a miracle*. Washington, DC: Counterpoint Press.

Burke, E. (1986). *Reflections on the revolution in France*. New York: Penguin. (Original work published 1790.)

Carson, R. (1984). *The sense of wonder*. New York: Harper & Row. (Original work published 1956.)

Colborn, T., Dumanoski, D., & Meyers, J. P. (1996). *Our stolen future*. New York: Dutton.

Commager, H. S. (1963). *Documents of American history*. New York: Appleton, Century, Crofts.

Critser, G. (2000). Let them eat fat. *Harpers, 300*(1798), 41–47.

Daly, H., & Costanza, R. (1992). Natural capital and sustainable development. *Conservation Biology, 6*(1), 37–46.

Earth Charter (2000). <http://www.earthcharter.org>.

Environment and Health Weekly, 2000. Nos. 681, 687, 688, 689.

Fishman, C. (1999). The smorgasbord generation. *American Demographics*, 54–60.

Gates, J. (2000). *Democracy at risk*. Cambridge, MA: Perseus.

Goleman, D. (1995). *Emotional intelligence*. New York: Bantam.

Greenspan, S. (1997). *The growth of the mind*. Reading, MA: Perseus.

Harman, A. (1998, September 30). Sad lonely world discovered in cyberspace. *The New York Times*, pp. 1, 22.

Healy, J. (1990). *Endangered minds*. New York: Simon & Schuster.

Healy, J. (1998). *Failure to connect*. New York: Simon & Schuster.

Horwitz, M. (1992). *The transformation of American law 1870–1960*. New York: Oxford.

Jefferson, T. (1975). *The portable Thomas Jefferson.* New York: Viking.

Joy, B. (2000, April). Why the future doesn't need us. *Wired*, pp. 238–262.

Kitman, J. (2000, March). The secret history of lead. *The Nation, 270*(11), 11–44.

Lasn, K., & Liacas, T. (2000, August–September). Corporate crackdown. *Adbusters Magazine*, pp. 38–49.

Leakey, R., & Lewin, R. (1995). *The sixth extinction.* New York: Doubleday.

Ledewitz, B. (1998). Establishing a federal constitutional right to a healthy environment in U.S. and our posterity. *Mississippi Law Journal, 68*(2), 565–673.

Living on earth (2000, May 21). National Public Radio.

Lockard, D. (1998). *Coal: A memoir and critique.* Charlottesville: University of Virginia.

Locke, J. (1965). *Two treatises of government.* New York: Mentor. (Original work published 1688.)

Matthews, R. (1995). *If men were angels.* Lawrence: University of Kansas.

M'Gonigle, M. (1999a). Ecological economics and political ecology. *Ecological Economics, 28*(1), 11–26.

M'Gonigle, M. (1999b). The political economy of precaution. In C. Raffensperger & J. Tickner (Eds.), *Protection public health and the environment.* Washington, DC: Island Press.

Nabhan, G., & Trimble, S. (1994). *The geography of childhood.* Boston: Beacon.

O'Brien, M. (2000). *Making better environmental decisions.* Cambridge, MA: MIT Press.

Postman, N. (1982). *The disappearance of childhood.* New York: Dell.

Postman, N. (2000). *Building a bridge to the eighteenth century.* New York: Knopf.

Press, E., & Washburn, J. (2000). The kept university. *The Atlantic Monthly, 285*(3), 39–54.

Putnam, R. (2000). *Bowling alone.* New York: Simon & Schuster.

Reisberg, L. (2000, June 16). Ten percent of students may spend too much time online. *Chronicle of Higher Education*, p. A43.

Scott, J. (1998). *Seeing like a state.* New Haven: Yale University Press.

Shepard, P. (1996). *Traces of an omnivore.* Washington, DC: Island Press.

Shepard, P. (1998). *Coming home to the pleistocene.* Washington, DC: Island Press.

Shrader-Frechette, K. (1993). Locke and limits on land ownership. In K. Caldwell, K. Lynton & K. Shrader-Frechette (Eds.), *Policy for land.* Lanham, MD: Rowman & Littlefield.

Shuman, M. (1998). *Going local.* New York: Free Press.

Sinsheimer, R. (1978). The presumptions of science. *Daedalus, 107*(2), 23–36.

Thornton, J. (2000). *Pandora's poison.* Cambridge, MA: MIT Press.

Wilson, E. O. (1984). *Biophilia.* Cambridge, MA: Harvard University Press.

Windle, P. (1994). The ecology of grief. *Orion, 13*(1), 16–19.

World Commission on Environment and Development. (1987). *Our common future.* New York: Oxford. (The Brundtland Commission Report.)

12

Eden in a Vacant Lot: Special Places, Species, and Kids in the Neighborhood of Life

Robert Michael Pyle

And *freedom* is this September field
Covered this far by tree shadows
Through which this child chooses to run
Until he chooses to stop. . . .
—Pattiann Rogers, "Concepts and Their Bodies (The Boy in the Field Alone)"

In a culture where growth and the advancement of civilizing forces against the frontier have always ruled, empty ground has represented little but an opportunity for more growth. The precarious condition of undeveloped land in modern cityscapes threatens the basic tie between humans and the rest of nature, as expressed and achieved by children in their preferred haunts. As the vacant lots go, so goes a source of intimacy and education that contemporary culture can scarcely afford to lose.

From the observed fact that a city abhors a vacuum, a peculiarly American vocabulary has arisen: *raw land* is a real estate term connoting acreage that is ripe and ready for conversion from a natural or undeveloped state to an "improved" condition; *waste ground* is acreage that may have had a productive use in the past but is now unused and therefore "wasted"; and perhaps the oldest and most widespread such term, *vacant lot*, is a tract, often platted or lying within a matrix of developed property, that remains ungraced by human structures and is therefore "vacant."

Unlike all these terms, which usually carry negative or at least unappreciative implications except in a speculative or opportunistic context, the phrase *open space* is generally applied positively, suggesting that the land in question has value for the very reason it *is* open. Yet open space may apply to the very same plots as raw ground, waste ground, and vacant lot. Recently I heard a hybrid moniker—*open ground*, which conveys all the ambivalence we as a colonizing culture have felt about such places.

Alone among these labels, *vacant lot* elicits widely affectionate following among one special group of users—children who have vacant lots in their lives. It is my premise here that nothing is less empty to a curious, exploring child than a vacant lot, nothing less wasted than waste ground, nothing more richly simmered in promise than raw ground. Yet too many adults seem to have forgotten the vacant lots of their growing-up years. In a word, the young and the grown often have different values about open ground.

In my classes and lectures, I make a point of asking the audience members or students whether they can remember a particular place where they made early contact with the land as boys and girls; a place they went repeatedly to play, explore, sulk, or think; a small, particular corner of the landscape where they went to make forts, catch creatures, and mess about with water and plants.

In most groups, most hands go up. I then ask them to picture the place and to tell me something about it. Commonly, the special spots are watercourses, such as creeks, canals, ravines, and ponds; a big tree, clump of brush, bosky dell, or hollow; parks, especially undeveloped ones; and old fields, pastures, and meadows. Very often, the term *vacant lot* is employed to define the place in question. In fact, insofar as most of my listeners have urban or suburban backgrounds, most of these sites of initiation are vacant lots of some sort. They share the qualities of nearness, wildness, secretiveness, and possibility. Most people can relate the details of the spot and tell stories from their places that surprise even themselves with their remarkable clarity and nuance and the deep affection aroused.

Next I ask a question whose answers tend to arouse feelings of both sadness and solidarity: How many can return to their special places and find them substantially intact?

A recent example brought a typical outcome. In October 1999, I addressed a group of managers, scientists, rangers, engineers, teachers, activists, and others involved in the future of the Cherry Creek Watershed in Colorado—the same watershed in which I grew up. When I asked the first question, almost every hand flew up. The participants' faces showed pleasure, excitement, and even reverie to be invited back into their childhood haunts—the very places that had lured them into their current professions and involvement with watersheds. But when I asked the second question, fewer than 10 hands went up in the audience of 300 or 400. Almost everyone's special place had been spoiled or greatly changed, and all those fallen faces told the story. As the following discussion showed, they took some relief in the fact that others shared their concern and distress over lost landscapes, and everyone emerged from the exercise bolstered in the common belief that such places must be saved wherever possible.

My own convictions along these lines came from a modest place indeed, an artificial watercourse that transects Cherry Creek on its run from the mountains to the plains. The High Line Canal emerges from the Platte Canyon in the Colorado Front Range and flows some 75 miles to its terminus in a small Platte tributary. A historic irrigation ditch that once was Denver's major water diversion, the canal now flows intermittantly and peters out near Denver International Airport. In an era of "epic liberties taken with water" (Reisner, 1986), when 16 major tunnels carry western slope water to the Front Range conurbation, the significance of the High Line has shifted from water movement to recreation. But in the 1950s and 1960s, "the ditch" played a massive role in my growing-up.

As denizens of the easternmost postwar subdivision on Denver's hems, an expansion of the small town of Aurora, we dwelt within a rigid grid of new streets laid down on farmed-out prairie. The Rockies attracted me mightily, but I lived on the Great Plains side of the city. From the day my older brother Tom came home and announced the discovery of a neat ditch full of intriguing wetness and greenery, the High Line Canal became my constant haunt, friend, and focus. Its tangled growth and sinuous path made the perfect getaway from the raw young suburb. Free to roam after school and in summer, I fled the town for the ditch every chance I got.

In the summer of 1954, Tom and I found ourselves caught out in a devastating hailstorm that broke the backs of cattle in the field next to us with hail the size of softballs. Small boys with no protection, we were literally in mortal danger. Tom swept me into a massive hollow cottonwood tree, and it saved our lives. Even so we were badly bruised, and he was concussed. Ever after, the hollow tree situated the center of the universe, and the canal took on even greater power.

When I left our ordered block, crossed an embryonic park, a set of tracks, U.S. 6, and a farmer's field or two to reach the canal, what I found was an unordered world of brown and green mystery. Long, broad-bladed grass hung over the banks and waved in the current. Chocolate wood nymphs flip-flopped among those grasses, big black-and-white admirals glided through the willows above, and still higher, their visual echo on the wing, the voluble, iridescent magpies. Orioles, flickers, kestrels, and kingbirds kept us constant company if we kept our slingshots holstered. I envied the few farm kids who actually lived along the canal and did my best to live what I imagined was their lives, while watching out for rough big boys and the ever-threatening farmers and ditch riders.

Over the coming years, I sought out the winding, cottonwood-shaded watercourse for purposes of exploration and play alone and with friends; discovery of crawdads, birds, and butterflies; sulking and kicking the dust through a troubled home life; hiding, camping, fort-building, stealing corn, cooking out, and pretending every kind of life in the out-of-doors; and ultimately, walking and parking and petting with girlfriends. But even in high school and later, when the chief social attractions lay in the city, I took myself to the fields and banks that had brought me up as a naturalist. I thought I was the only one, but I was wrong. Of my few peers who later became involved in life sciences—a vet, a zookeeper, a federal wildlife agent—all attributed their interest at least in part to the High Line Canal. And so I believe it goes, for such places everywhere.

I have told the story of the High Line Canal and its impact on myself and others in *The Thunder Tree: Lessons from an Urban Wildland* (Pyle, 1993). Attempting to capture the canal's overall import, I wrote (p. 152): "Had it not been for the High Line Canal, the vacant lots I knew, the scruffy park, I'm not at all certain I would have become a biologist. . . . The total immersion in nature that I found in my special spots baptized me in a faith that never wavered, but it was a matter of happenstance, too. It was the place that made me."

What is most germane here is the way in which the canal changed as Aurora, Colorado, grew from 40,000 to some 400,000 people. Fields, marshes, farmyards, and woodsy gulches all went the way of the D-9 Caterpillar blade. It is true that these rampant losses first inspired me toward conservation activism; it is also true that the efforts of many who loved the canal and what it represented brought about eventual protection for a few sites and consecration of the ditch riders' road as a part of the National Trail System. But much of the charm and fundamental fascination of the place were lost, and as they went, the canal's ability to influence the lives of children diminished. If such changes matter at the High Line Canal in Colorado, then the accumulated affect of millions of special places lost all across the continent must be dramatic.

The intense connection that children form with particular places is no secret. Nixon (1997) gives a useful summary of psychological research in this area. Kahn (1999) investigated the origins of children's connections to the natural world through analysis of both physical and social sciences, concluding that these bonds are fragile and subject to stresses peculiar to contemporary culture. Nabhan and Trimble (1994), Trimble (2000), Sobel (1993), and Stafford (1986) all explored "the geography of childhood" (Nabhan and Trimble's title), including detailed descriptions of their own places of fort-making, escape, and discovery. Sobel's "labyrinthine passages in the rhododendrons" and Stafford's "path threading through nettle and alder" make it clear that these scholars and writers owe their naturalist lifeways to their hearts' own places.

The recognition of the importance of outdoor experiences for the young greatly predates these contemporary thinkers, however. From the late nineteenth century to the mid-twentieth century, nature study, as a formal element of school curricula, overlaid this bush savvy with actual pedagogy for many American children. The quality and depth of such instruction varied widely, of course. But most American schools employed Anna Botsford Comstock's *Handbook of Nature Study* (Comstock, 1911), and I have collected dozens of less ambitious but similar texts from the same period. While the nature study movement certainly did not produce a nation of sophisticated naturalists, it did ensure that most students received at least a framework of tutelage on which to hang their own outdoor discoveries.

The goal of accelerated mathematics and science instruction for all students both before and after World War II, peaking during the cold war, all but extinguished natural history as a standard curriculum unit. Hard, numerical sciences took its place, and the observational approach fell into disrepute. Even the term *naturalist*, revered in John Burroughs's day, became something of a derogation, implying a soft, nonscientific, and even sentimental approach toward an abstract nature. Never mind that descriptive natural history and systematics could be conducted with as much rigor as outright experimentation; they quickly became passé and lost support in the schools.

My own intense interest in natural history was certainly not accommodated at school, except by a few exceptional teacher-birders who offered a saving after-school ecology club. I was obliged to make do with books (including those dusty old nature study texts, which I began collecting in junior high school), National Audubon Society pamphlets, and truly self-guided walks. Nowadays, children may have formal environmental education to augment the classroom. Sophisticated in some respects, employing water-quality tests and other modern techniques with which Anna Botsford Comstock was not familiar, E.E. nonetheless fails to replace the forsaken nature study of earlier times. True, some innovative teachers employ map-making, tracking, and other field-based exercises. But outdoor education's common emphasis on games, teamwork, and motion militates against deliberate and ruminative natural history. Kids get lessons on ecological relationships, without much basic knowledge of the related organisms and their lifeways—not even, or especially not, their names.

Few students (or teachers) have even the most basic acquaintance with their local fauna and flora when they graduate. As Orr (1994, p. 126) put it, "Even in this time of ecological concern, high schools, colleges, and universities continue to turn out a large percentage of graduates who have no clue how their personal prospects are intertwined with the vital signs of the earth." Nor is ecological illiteracy limited to the cities. As a friend of mine from a rural logging and farming family recently put it, "I cannot believe the people who have lived in this area all their life long who do not know a salamander from a salmonberry." If this is true in the country, how much more critical must our ignorance be in the cities? This is not an academic question, as the population concentrates more and more away from the countryside.

In modern times, the special places sought out by the young have furnished an alternative education parallel to the classroom. All children are autodidacts, their subjects depending on the available curricula: if it is the streets, they will become street-wise; if it's computers, they'll be screen wizards. But if what is at hand is a scrap of the wild, at least some children will become naturalists before ever receiving instruction. Though this natural tradition is intuitive and associative rather than rote and employs self-generated vernacular nomenclature (where I live, water striders are "skippers," and cinnabar moth larvae are "tansy tigers"), it triggers not only knowledge but also intimacy with nature among the young.

As cities began to grow in earnest, the demise of traditional field studies placed an increased burden on open spaces: if kids were going to become acquainted with nature, it would have to take place mostly out-of-doors, on their own time. Yet even as nature study faded, eclipsed in the shadow of *Sputnik*, the postwar housing spurt claimed woods and fields at a rate never before seen. And even as increasing numbers of families left the farmlands for the cities, the cities grew out to subsume their farmland edges. And as special places retreated before the growth of the suburbs and the towns moved to or became the cities, many children were left in little physical contact with the world outside their homes, schools, and shopping malls.

To begin to imagine the consequences of the destruction of urban open ground frequented by children, we must first consider the actual services such landscapes provide. The first that comes to mind is *play*. For much of our history, when children have been left to their own devices, their first choice has often been to flee to the nearest wild place—whether a big tree or brushy corner in the yard or a watercourse or woodland farther away. This is where they can imagine and enact adventure, construct forts and intrigues, and hunt crawdads and bugs. In aboriginal societies, this kind of play was essential for forming basic survival skills: today's crayfish and minnows are tomorrow's game and pot-fish. As the needs for bush skills evaporated, the atavistic pleasure in such play did not, and it continues to connect us to our hunting and gathering past, to our evolutionary legacy.

Second, I think of nature literacy. In tribal times, a close knowledge of other species was essential to survival. Most people grew up knowing many of their nonhuman neighbors, or if they didn't, they were in trouble. The argument can be made that because survival is no longer dependent on the individual's sophistication in the ways of the wild, we need not bother to foster it. My response would be that collective ignorance leads inexorably to collective indifference; and from there, it is not many more steps to ecological depreciation and collapse. I have argued (Pyle, 2000b, p. 96) that "Maybe excepting only human population and acute chemical pollution, the greatest threat to a sensible environmental future is nature illiteracy."

The third and perhaps most important service rendered by the vacant lots, creeks, and back forties is literacy's partner, intimacy. Widespread public intimacy with the natural elements of our surrounds gives an opportunity to avoid the extinction of experience (Pyle, 1978, 1993). The extinction of experience postulates that daily contact with a diversity of experiences—botanical, zoological, cultural, architectural, social—leads to an appreciation for the elements of a rich setting, if only implicit; to concern; and ultimately to the desire to conserve. But when common species disappear, they might as well (in one sense) be wholly extinct for those with a narrow radius of reach. Further, when variety thus drops out of the local scene and a sameness sets in, the sequelae may include widespread ignorance of the world's diversity, followed by disaffection, alienation, and apathy. These are not the states of mind that inspire conservation activism. So local extinctions continue, leading to still flatter and more depauperate environments and still deeper isolation from richness. The extinction of experience is thus a cycle whereby impoverishment begets greater impoverishment.

The ultimate result of the extinction of experience might resemble a superurban condition described by Ashley Montagu and Samuel Snyder (1972, p. 198) in *Man and the Computer*: "The city has become a wilderness in which human beings lose their humanity because inhumanity has become the way of life of the city—indifference, disengagement, and isolation." It is just this state of disengagement that makes the extinction of experience such a threat today.

A broad social construct like the extinction of experience is difficult to test analytically, consensus on its reality lending itself more, perhaps, to a common recognition of shared experience than to controlled experimentation. However, a strongly suggestive indication of the degree of natural change (which I call ΔN) can be apprehended by examining the response of the butterfly fauna to change in a given locality. Butterflies, because they are often ecologically specific, conspicuous, popular, and well known, have been used increasingly as effective indicators of ecological and biogeographical richness, uniqueness, and health (Pyle, 1976). As habitats are developed, homogenized, and fragmented, numbers of individuals and of species decline (New, Pyle, Thomas, Thomas & Hammond, 1995). Furthermore, ΔN derived from local butterfly extinctions can be read as a measure of the suitability of habitat to influence the young, as it is often the insects, and in particular the butterflies, that arouse the curiosity of children, as expressed in the making of collections and observations (Ehrlich & Ehrlich, 1981; Wilson, 1994).

I have studied the butterfly fauna along the High Line Canal, intermittently, from 1959 to the present. Over 40 years I have observed and documented changes—some additive, most subtractive. Annual butterfly counts for the past 25 years (Chu, 2000) have added to this database. By 1970, I had recorded one-tenth of the North American butterfly fauna along the High Line Canal (Pyle, 1974). In subsequent years, when I accounted for apparent and imminent local extinctions in my study area, I found that the Auroran extinction rate (40 percent) was higher than that calculated for Staten Island (23.6 percent), Orange County, California (21.25 percent), or San Francisco (7.6 percent) (Pyle, 1983). Since then, a few of the species have reappeared, others have dropped out, while new species have been recorded. But chiefly, the additions consist of erratic, adventitious irruptives and immigrant species and generalists, unreliable from year to year; while colonial habitat specialists have dropped out. For example, the High Line Canal Fourth of July butterfly count has recorded the national high for the introduced European cabbage white (*Pieris rapae*), a great generalist, for several of the 25 count years (Chu, 2000); whereas a colony of the related, specialized Olympia marble (*Euchloe olympia*), formerly abundant in certain places, has vanished (Pyle, 1974).

If we further characterize ΔN by considering the area in hectares (A) and the time (T) over which the extinctions (E) were incurred, such that $\Delta N = \%E/A/T \times 1{,}000$, then a measure of the intensity of loss can be gained. By this formula, a nature reserve of 1,000 hectares that loses one species out of a fauna of 100 through stochastic processes over 100 years would have a ΔN value of 0.01. The four urban areas considered above would have ΔN values of roughly 0.11 for Staten Island, 0.14 for Orange County, .80 for San Francisco, and 143.0 for the High Line Canal in Aurora.

Since the area may relate more to biological consequences (e.g., island extinction theory) than cultural, it could be excluded from the equation. Doing so flattens the numbers somewhat. Calculating and including the radius of reach for particular human populations, or selecting equal-sized sections of land to compare, would also change the final numbers. However, the relative order remains the same. And even allowing for a large margin of error, these figures suggest that *the opportunity for experiencing nature in the cities studied has diminished measurably, most of all in the most rapidly suburbanizing of them.*

Furthermore, the homogenization of urban faunas and floras proceeds largely through the loss of small, marginal remnants of formerly widespread habitats. Insects are able to persist under these conditions longer than megafauna: the last habitats for both the Olympia marble and the bronze copper (*Lycaena thoe*) in Aurora were, respectively, an old field and a marsh fragment in a vacant lot (Pyle, 1993). And since urban children obtain most of their natural contact in exactly these kinds of places, we may conclude that their opportunities for natural discovery decline in proportion to urban butterfly extinctions.

Finally, factoring in the high proportion of biologists, conservationists, and other resource professionals who gained their initial inspiration through contact with insects, ΔN as indicated by urban butterfly extinction may be considered a reliable index of the extinction of experience in action and a predictor of its effects. While it is not possible to demonstrate that environmental leaders will no longer arise from the streets of contemporary Aurora, it seems reasonable to conclude that the likelihood that young Aurorans will become avid naturalists and conservation voters has been reduced in parallel with the town's butterflies and the diversity they represent.

An important corollary of the extinction of experience holds that its effects operate especially strongly on those whose *radius of reach* is small—people who cannot easily reach beyond their neighborhoods for stimulation. This means, among others, the poor, the disabled, the very old, and the young. Future Auduboners emerging from the urbs are likely to be those children whose families can afford to send them away to camp or to take frequent sallies to the countryside—or those with a special place within walking or biking distance from home. When what you see in the 'hood is all you get, the extinction of experience can have especially baleful consequences.

These consequences go beyond the initial cycle of the extinction of experience. When experiential contact with nature, in the broadest sense, is diminished, negative impacts spread out to every cultural level. *Physically*, youngsters suffer from the absence of exercise in fresh air that outdoor scrounging traditionally provides. The current epidemic of early onset obesity owes much to sedentary substitutes for outdoor play. This too creates a feedback loop, for the more slothful a child grows, the less likely he or she will be to seek physical play beyond the television or computer room, and the more likely he or she will habituate to recreational, even compulsive, eating.

We can further infer that nature-deprived children suffer *intellectually*, at least on some levels. Clearly, mental development does not wholly depend on abundant contact with plants, animals, soils, and rocks. Many people achieve keen intellectual powers in largely indoor settings. However, I would argue that breadth of awareness, facility of reasoning, acuity of observation, and the kinds of associative skills that enhance cerebration may all sharpen as a direct result of biological and geological exposure. Support for this and similar conclusions appears throughout the growing literature on biophilia, from E. O. Wilson's original expression of the hypothesis (Wilson, 1986) to a gathering of essays on the subject (Kellert & Wilson, 1993). In a recent statement on the cultural effect of biophilia, Kellert (1997) maintains that human powers of creativity and imagination, as well as our emotional well-being, owe much to the nonmanufactured world.

If Kellert, Wilson, and others are right, then the places that arouse biophilia must affect us *emotionally* as well. The young people I know whose lives are rich in natural experience seem to be on firmer, better balanced footing—and to be happier—than those whose universes orbit strictly around people and made things. Besides, as I know from my own experience and an abundance of testimony, the pain of losing one's special place is a deep and abiding betrayal that never fully recedes. It follows that preventing such losses will head off great hurt. In more positive terms, those on intimate terms with wild places find the world emotionally enriching.

It is no large stretch, then, to imagine that the loss of contact can be *morally* depleting. We're not talking about rearing a generation of Saint Francises, and it is true that much outdoor play (especially among boys) involves testing cruelty and destruction. Nor is it the case that naturalists are good and urbanists are bad, ipso facto. But there is arguably a moral and ethical dimension that emanates or prospers in the light of personal immersion in the wholeness of the physical world. After years of study of human attitudes toward other life forms, Kellert (1996, p. 218) concluded that "the more we plumb the depths of nature, the more we encounter its unrivaled capacity to nurture the human body and spirit."

The moral element has also to do with the maturation of ethics to extend beyond the strictly human realm. Cogent propositions of a moral relationship with the natural world may be found in Aldo Leopold's "Land Ethic" (Leopold, 1949), William O. Douglas's *Wilderness Bill of Rights* (Douglas, 1965), Roderick Nash's *The Rights of Nature* (Nash, 1989), and Christopher Stone's "Should Trees Have Standing?" (Stone, 1972). Each of these thinkers makes a clear case for an advanced moral universe that takes into account the extrahuman. Surely such an ethic would be more likely to develop in a nation that preserved its points of free and easy contact with rocks, trees, and wilderness—or in Leopold's inclusive term, the land, in small parcels as well as large.

Depletion of our natural context, both in quantity (or time) and quality (or depth) of exposure, may interfere with the development of these and other human qualities. As Montagu and Snyder (1972, p. 199) wrote, urban "separation from nature leads to a view of it that is wholly disengaged, even alienated, and frequently hostile. This is a pathological state, a morbid dissociation from what should always have remained a vital involvement with nature." Conversely, everything that acts to reconnect human culture with nonhuman nature counteracts the pathology. But in a macabre extension of the theory, the increase of sociopathic behavior actually accelerates the extinction of experience. Sadly, another factor must now be considered when it comes to children and nature—their personal safety.

Cuts, scrapes, and broken bones as unwonted trophies of outdoor adventures have always been with us. But now the panoply of threats has expanded to include abduction and personal harm at the hands of adults. These specters are not new but were so rare in former times as to represent nothing more than a cautionary bogey. With population expansion and crowding, the frequency of assaults—or its perception—has increased to the point that few parents are comfortable allowing their children anything like the outdoor freedom and latitude that my generation took not only for granted but as an essential birthright. Whether or not violent incidents represent a genuine danger or one largely projected through sensational news coverage, parents *think* that the woods are unsafe. Recently, I met a woman, an academic sensitive to the natural world, who nonetheless would not permit her seven-year-old outside the family's cul-de-sac unaccompanied. This would have been sheer torture to my friends and me. I feel strongly that had I not enjoyed the freedom of the canal, I would not have become who I am. The loss of footlooseness among the young must be counted every bit as much a tragedy and a challenge as the loss of places in which to be footloose.

Now, in addition to the absence of formal nature study, the retreat of diverse habitats from the home ranges of the young, and the shrinkage of those ranges due to security concerns, we are experiencing the rise of the virtual in place of the real. Television has long commanded much of children's discretionary time: *The Mickey Mouse Club* and *Spin and Marty* certainly competed with the backyard for me, more than 40 years ago, when only three TV stations were available in Denver. But in the past 10 years, computer games, the Internet, and other forms of secondhand entertainment have come to occupy an ever more enormous portion of childhood's hours. It is easy to poke fun at the man I saw speaking on his cell phone in a butterfly house in Seattle, ignoring the vast blue morphos shimmering all around his head in favor of his current wireless crisis; after all, electronic narcissism is everywhere on the street corners and freeways. But I was disturbed to see two brothers buried in their Gameboys as our 777 passed over the sunlit and unobstructed mountains, icefields, and calving glaciers of Greenland and, more so, to have the flight attendant ask us to lower our window shades, the better to see the video screens.

I have also found disquieting a rash of technical papers describing the boundless wonders of "class field trips in cyberspace." One such article states: "Increasingly, technology is being used by educators to take students on 'virtual' excursions to a variety of fascinating places" (Holzberg, 1997, p. 42). I know that some of these electronic field trips furnish elegant and participatory instruction. "Journey North," for example, allows students anywhere to follow the migrations of monarch butterflies and birds, as well as flowering rates, throughout North America, as they contribute their own observations. A videodisc is available that takes students "afield" with Lincoln Brower and his helpers studying the field ecology of monarchs. Used together, these could be powerful tools for understanding a subtle and remarkable organism and its world—especially if the kids get outdoors to see actual monarchs in an actual milkweed patch. There is the key: the electronic element should serve as a spur and a link for real-world watching, rather than a substitute. But not all of these media include the field component. At the same convention of science teachers where I was buoyed by an excellent exhibit of a hands-on program called "Kids in Creeks" sponsored by the Bonneville Power Administration, teachers spoke of principals content to let CD-ROMs, hypercards, and the Internet take the place of messy field trips with their insurance and staffing problems.

Of course, electronic mediation pertaining to natural history themes may effectively convey facts and impressions and generally reinforce interest in animals and geography. But when the world comes edited for maximum impact and bundled into quick bites and bytes, it fails to convey the everyday wonder of the much-maligned ordinary. Just as real life does not consist primarily of car chases and exploding buildings, quotidian nature is much more about grasshoppers in the pigweed than it is rhinos mating on a pixilated screen. Even butterfly houses and zoos, though they present a kind of primary experience, bear some culpability here. Displays of extravagant animals behaving dramatically in captivity and on television can spoil the young for the real thing outside their door. Those big, brilliant blue morphos in the tropical house should not excite more fascination than the equally bright but tiny spring azures among the dogwoods by the ditch.

But what ditch? That is the rub. For when the ditches and creeks are all sent down the storm drains, concreted, or sprayed, they are lost to the children. Lucky is the child of the city or suburbs who still has a richly inhabited ditch, creek, field, or forest within walking distance of home. Nor, I emphasize, do parks and nature reserves make up for what I call the secondhand lands or hand-me-down habitats, which correspond to what British naturalist Richard Mabey (1973) describes as the "unofficial countryside." Parks are normally too manicured and chemically treated to offer much of interest to the adventuring youngster. And as for nature reserves, they might as well be paved over for all they offer in the way of boundless exploration. For special places to work their magic on kids, they need to be able to do some clamber and damage. They need to be free to climb trees, muck about, catch things, and get wet—above all, to leave the trail. Such activities are normally proscribed in reserves and for good reason. I support the strict protection of natural areas wherever possible, for the careful perpetuation and management of scarce elements of diversity. But the unofficial countryside—the domain of unsupervised outdoor play—needs to be recognized and protected among the built landscape, as well as the official preserves. As I put it in *The Thunder Tree*, "We need to recognize the humble places where this alchemy occurs, and treat them as well as we treat our parks and preserves—or better, with less interference" (Pyle, 1993, p. XIX).

In making such assertions, I must also recognize the assumptions they depend on. One of them is the actual extent of vacant spaces. It is possible to cruise timber to determine board feet of stumpage but much more difficult to measure the current availability of potentially special spaces for children or the amount and quality of open land that is necessary to ensure the kind of child-habitat encounters that allow them to develop caring attitudes. Furthermore, we do not know with precision the degree to which vicarious and virtual experiences with nature may compensate for direct contact or the depth of impact when contact is thwarted. To some extent we are dependent here on our deductive powers and considered speculation operating on anecdotal observation. This is an important caveat when considering my conclusions. But until robust data become available, it seems reasonable to assume that rapidly expanding suburban development and experiential vicariance intensify the extinction of experience cycle.

If this is so, what about practical antidotes and alternatives to the loss of special places? For the children have not been cast out of their edens: their edens have been pulled out from under them. How do we put them back and keep the ones that remain?

Maybe the greatest challenge is to identify and protect the special places. Since secrecy is one of their hallmarks, such places are seldom well known by adults. They also tend to be humbler than the kinds of places conservationists fight for. But even when the vacant lots and open ground beloved of children do enter the land-use discourse, they seldom command the power or resources to save them. Typically, the development value of the site is so great as to simply sweep away all other concerns. Rare success stories usually involve major private donors, the presence of endangered species, or some sort of "miracle" of cooperation or progressive planning.

A classic allegory of last-minute salvation presented as a children's story may be found in George Selden's *Tucker's Countryside* (1969). In this modern fable, the children save the meadow from the developer by shaming the chairman of the town council with the "discovery" of the founder's homestead, "a matter of the deepest historical significance," as he then blusters, before pronouncing that it will be left as it is "as a natural shrine in memory of the great pioneer." But the children prevail only with the assistance of the animals, who help them plant bogus evidence leading to the historic designation. In reality, the magic element is generally missing, and the political imperatives are less malleable to the people's will.

Saving special places as individual *causes célèbres* can sometimes succeed, but it is tedious, expensive, and risky. And even when the children seem to prevail, the ultimate outcome can be undermined by hard realities. Not long ago, a relictual Valhalla not far south of the High Line Canal in Aurora, Colorado—one of the last undeveloped tracts for miles around—came under threat of development. Many considered the replacement of Jewell Wetlands by hundreds of condominiums inevitable. But local parents and others organized, raised the money in partnership with the city, the drainage district, and the state lottery, with assistance from the Trust for Public Land. In a stunning victory, they bought the prairie remnant, cattail marsh, and its willow-and-cottonwood fringe for open space and flood control. The children whose after-school and summer habitat it was celebrated to know that they would continue to explore and play and learn at Jewell. And so they shall. However, the many interests involved, the prior damage from off-road vehicles, vandals, and filling, the unstable gully and other perceived dangers, and the diverse clientele for the new park led to extensive changes. Volunteers by the hundreds helped landscape a handsome and functional place combining some natural grassland and marsh with drainage and park features. The Aurora Jewell Wetlands were heroically saved for the pleasure of the many. But in the process, the wild tangle and its invitation to uncontrolled discovery were largely lost. Now, boardwalks and butterfly gardens replace sweet surprise, and the creek falls in concrete runnels. When tradeoffs prevail, sacrifice happens.

How much better if planners incorporated retention of unmanicured open spaces in their urban growth master plans. Unfortunately and ironically, exactly the opposite often occurs, for a conservation rationale: to prevent sprawl at the edges of towns the "New Urbanism" has embraced the concept of *infilling* with gusto—maximizing the density of development inside of cities instead of allowing it to dribble out the edges as always before. While infilling may help to maintain city limits, it is anathema for the lovers of vacant lots and "waste ground." This conundrum is currently the subject of intense debate in Portland, Oregon, a city that is considered very green but that also pushes to grow beyond its legislated urban growth boundary. Though it is considered a naughty word in Portland, this conflict is nothing else but an expression of overpopulation: just too many people living in one place to maintain what they all came for.

A partial saving grace in Portland is the vigorous effort by urban conservationists to build on the city's original Olmsted Brothers plan, which included a 40-mile loop of green space. While the loop was never finished and many gaps exist, Portland visionaries have expanded it to a 140-mile linked loop of habitat nodes and corridors. If they are successful, they may counteract the negative effects of infilling to a meaningful degree. But corridors by themselves will not save every child's special place. They work only for children who happen to live nearby; for unlike birds, kids need habitat right there, underfoot.

The creation of corridors can be a cost- and ecology-effective technique for maintaining the connection between cities and their natural underpinnings. But they work best for children if they are linked to intersecting, concentric, tangential, or parallel paths and if they are studded with many large and small clots of unimproved habitat. Just as corridors linking archipelagoes of habitat islands support more species than isolated patches can do, they also have the attribute of expanding available urban wildland beyond the actual acreage of the ground involved. This method thus has the power to extend the radius of reach of children farther and farther as their powers of wandering increase with age. The High Line Canal offered what seemed unlimited scope for exploration to a child of the 1950s, though it was all technically off-limits. Even as the habitat along it was built up, a public greenway of some 60 miles was developed, eventually connecting with other trail systems, creeks, and gullies crossed by the canal. Thus today the Canal trail offers a greater extent of access for children than we ever had, even if its amplitude for natural discovery has been much reduced. The diversity too could have been maintained if more habitat nodes had been saved.

In the 1970s, I lived in Seattle and became acquainted with green places that my mother had known as genuinely wild many years before. I imagined a 12-mile "green circle" that would link these through the severed and frayed remnants of another uncompleted Olmsted plan. The "green leaks" grew, however, along with Seattle and the University of Washington. One bright light in this diminishing scene has been a fervent citizen campaign to daylight Ravenna Creek, which runs out of the same Ravenna ravine my mother haunted as a girl and empties into the vestigial marsh at the margin of Union Bay, my naturalist's retreat when at university. An upscale shopping center stymied the original plan for the daylighted stream, but other routes have been plotted, and the momentum gives hope for eventual success. What could mean more for children's experience than bringing back the creeks they once lost?

Nor are vacant lots an irreplaceable resource. In a presentation to the Natural Areas Association meeting in St. Louis in October 2000, Yale forest sociologist William Burch discussed opportunities inherent in urban problems. Due to downtown decay, Detroit is presently plagued—or blessed—with some 30,000 vacant lots. "We're creating a national forest in inner-city Detroit," Burch quipped. This statistic is seen as a large negative by urban advocates, but I too find it exciting: often the first to suffer from the extinction of experience, perhaps the inner city will also be the first to discover the treasure in abandoned land.

Since restoration of built and paved habitats is feasible, and since many urban organisms (even semispecialists) may recolonize restored sites or be reintroduced, extinction of experience need not necessarily be forever (Pyle, 2000a). Frayed and severed connections can be rewoven. And we must remember that children themselves are powerfully adaptive organisms: for them a washing machine box can be a cabin, or a thicket can become a jungle. Children are masters at making mountains out of molehills and trackless veldt from a prairie dog town in a farmed-out field. But they are also connoisseurs of surprise, mystery, and wildness writ small. The planner who attempts to plug in a structured playground or park in place of a scruffy canebrake or a sumac patch with a pond will fail.

Our cities need to maintain the natural habitats of children—undedicated, unmanaged, undeveloped ground where unplanned, unsupervised, and unexpected discovery can take place. Realtors will continue to call it "waste ground." But in my view, nothing is less wasted than ground where the hand of man has held back and the minds of boys and girls can engage with plants and animals and dirt, nothing more sacred than land that is yet raw and ripe with surprise.

Equally, we need to retrain the impulse to "improve" open land into developed parkland. When children go down to the woods and fields, will their experience be signed, led, guided, planned, and programmed? Or will it be a spontaneous exercise of connection between the land, the water, and their own imaginations, with all the chance in the world to simply be, with nothing between themselves and epiphany but the length of the day?

It may be that, given the demographics and politics of contemporary America, such expectations are simply unreasonable. After all, the tamed experience of nature (though inferior, in my view) is better than none. Maybe the only way to "save" such land from off-road vehicles, vandals, developers, and other rascals is to tame it. Maybe. But to a curious kid, wildness not replaceable by artifice, and kids who encounter wildness may become people who care.

Children know what is interesting when they see it. What I fear is that in the absence of intriguing places to prowl and in the presence of a growing peer obsession with gadgets, they will forget to remember what it is.

We who can remember our own creeks and vacant lots should recognize all they represent. In our earnest desire to maintain our own children's edens, we must remember that children everywhere, not just those of privilege, have such a need. To begin to reverse the loss of special places, we have no choice but to admit the primacy of untended ground in our cultural landscape. And once we accept that, we must take the following direct actions on behalf of the vacant lots and their like:

- Ask our kids to tell us about their special places, maybe even show us.
- Inventory the unofficial countryside within a mile's walk of our homes and in other neighborhoods.
- Seek to influence our planners and decision makers to take into account the extraordinary occupation rate of vacant lots, the infinite investment value of waste ground.
- Fight doggedly for these places whenever the threat arises, by all available legal means, until developers and commissioners realize that it won't be easy to wreck them.

• Bully parks departments into leaving well enough alone when it comes to wildland "improvements."

• Organize parents and older siblings to be on hand for exploration sessions—not *with* the children, neither leading nor organizing and definitely not supervising, but close enough to satisfy themselves that the kids are safe.

• Limit the amount of time spent indoors in front of the television, the computer, and even books.

• Encourage nature reserves and environmental education programs, but do not allow them to take the place of spontaneous adventuring on rough ground.

• Build constituencies and partnerships among nonprofits and agencies to imagine and assemble kid-habitat corridors and green circles richly studded with habitat nodes and to plug the green leaks as opportunities arise.

• Work with naturalists to characterize the species-rich sectors remaining in our towns and neighborhoods, for these will be the same places where vital experience may still be found.

We must do all this, and above all we must remember Calvin's pronouncement to Hobbes, after the comic-strip lad and his tiger had suffered a particularly trying day: "The world's not such a bad place," says Calvin, "when you can get out *in* it."

References

Chu, J. (2000). High Line Canal, CO (23 years). *1999 Report, NABA/Xerces Fourth of July butterfly counts.* North American Butterfly Association. Morristown, NJ.

Comstock, A. B. (1911). *Handbook of nature study.* Ithaca, NY: Comstock.

Douglas, W. O. (1965). *A wilderness bill of rights.* Boston: Little, Brown.

Ehrlich, P., & Ehrlich, A. (1981). *Extinction: The causes and consequences of the disappearance of species.* New York: Random House.

Holzberg, C. S. (1997). Little kids, big projects. *Technology and Learning, 18*(3), 42.

Kahn, Peter H., Jr. (1999). *The human relationship with nature: Development and culture.* Cambridge, MA: MIT Press.

Kellert, S. R. (1996). *The value of life: Biological diversity and humans.* Washington, DC: Island Press.

Kellert, S. R. (1997). *Kinship to mastery: Biophilia in human evolution and development.* Washington, DC: Island Press.

Kellert, S. R., & Wilson, E. O. (Eds.). (1993). *The biophilia hypothesis.* Washington, DC: Island Press.

Leopold, A. (1949). The land ethic. In *A Sand County almanac and sketches here and there.* New York: Oxford University Press.

Mabey, R. (1973). *The unofficial countryside.* London: Collins.

Montagu, A., & Snyder, S. (1972). *Man and the computer.* Philadelphia: Auerbach.

Nabhan, G. P., & Trimble. S. (1994). *The geography of childhood: Why children need wild places.* Boston: Beacon Press.

Nash, R. (1989). *The rights of nature: A history of environmental ethics.* Madison: University of Wisconsin Press.

New, T., Pyle, R. M., Thomas, J. A., Thomas, C. D., & Hammond, P. C. (1995). Butterfly conservation management. *Annual Review of Entomology, 40,* 57–83.

Nixon, W. (1997). How nature shapes childhood. *Amicus Journal, 19*(2) 31–35.

Orr, D. W. (1994). *Earth in mind: On education, environment, and the human prospect.* Washington, DC: Island Press.

Pyle, R. M. (1974). The butterflies of the High Line Canal of Colorado. *Mid-Continent Lepidoptera Series,* 2(24), 1–19.

Pyle, R. M. (1976). *The eco-geographic basis for Lepidoptera conservation.* Ann Arbor: University Microfilms.

Pyle, R. M. (1978). The extinction of experience. *Horticulture, 56*(1), 64–67.

Pyle, R. M. (1983). Urbanization and endangered insect populations. In G. W. Frankie & C. S. Koehler (Eds.), *Urban entomology: Interdisciplinary perspectives.* New York: Prager.

Pyle, R. M. (1993). *The thunder tree: Lessons from an urban wildland.* Boston: Houghton Mifflin.

Pyle, R. M. (2000a). Resurrection ecology: Bring back the Xerces blue! *Wild Earth, 10*(3), 30–34.

Pyle, R. M. (2000b). *Walking the high ridge: Life as field trip.* Minneapolis: Milkweed.

Reisner, M. (1986). *Cadillac desert: The American West and its disappearing water.* New York: Viking.

Selden, G. (1969). *Tucker's countryside.* New York: Camelot/Avon.

Sobel, D. (1993). *Children's special places: Exploring the role of forts, dens, and bush houses in middle childhood.* Tucson: Zephyr Press.

Stafford, K. (1986). A separate hearth. In *Having everything right*. Lewiston, ID: Confluence Press.

Stone, C. (1972). *Should trees have standing?* Los Altos, CA: William Kaufman.

Trimble, Stephen. (2000). Untitled. In E. Goodenough (Ed.), Secret spaces of childhood (pt. 2). *Michigan Quarterly Review, 33*(3), 521.

Wilson, E. O. (1986). *Biophilia: The human bond with other species*. Cambridge, MA: Harvard University Press.

Wilson, E. O. (1994). *Naturalist*. Washington, DC: Island Press.

Contributors

Louise Chawla
Whitney Young College
Kentucky State University
Frankfort, KY 40601
chawla393@aol.com

John D. Coley
Department of Psychology
Northeastern University
MS 0125 NI
360 Huntington Avenue
Boston, MA 02115-5000
jcoley@neu.edu

Frans B. M. de Waal
Yerkes Regional Primate Research Center
Emory University
954 North Gatewood Road
Atlanta, Georgia 30329
dewaal@emory.edu

Judith H. Heerwagen
J. H. Heerwagen & Associates Inc.
2716 NE 91st Street
Seattle, WA 98115
j.heerwagen@worldnet.att.net

Peter H. Kahn, Jr.
Department of Psychology
University of Washington
Box 351525
Seattle, WA 98195-1525
pkahn@u.washington.edu

Rachel Kaplan
School of Natural Resources and Environment
University of Michigan
430 East University
Ann Arbor, MI 48109-1115
rkaplan@umich.edu

Stephen Kaplan
Department of Psychology
University of Michigan
525 East University
Ann Arbor, MI 48109-1109
skap@umich.edu

Aaron Katcher
University of Pennsylvania
PAN-ATA
3008 Spring Oak Place
Arlington, TX 76017
akfalco@aol.com

Stephen R. Kellert
School of Forestry and Environmental Studies
Yale University
Sage Hall
205 Prospect Street
New Haven, CT 06511
stephen.kellert@yale.edu

Olin Eugene Myers, Jr.
Huxley College of Environmental Studies

Western Washington University
Bellingham, WA 98225-9085
gmyers@cc.wwu.edu

Gordon H. Orians
Department of Zoology
Box 351800
University of Washington
Seattle, WA 98195
blackbrd@u.washington.edu

David W. Orr
Environmental Studies Program
Oberlin College
The Lewis Center
Oberlin, Ohio 44074-1095
david.orr@oberlin.edu

Robert Michael Pyle
369 Loop Road
Gray's River, WA 98621-9702
tlpyle@willapabay.org

Carol D. Saunders
Communications Research
Brookfield Zoo
3300 Golf Road
Brookfield, IL 60513
casaunde@brookfieldzoo.org

Patrick Shafto
Department of Psychology
Northeastern University
MS 0125 NI
360 Huntington Avenue
Boston, MA 02115-5000
shafto@neu.edu

Gregg E. A. Solomon
Department of Brain and Cognitive Sciences
Massachusetts Institute of Technology
NE20-437
77 Massachusetts Avenue
Cambridge, MA 02139
geas@psyche.mit.edu

Cynthia Thomashow
Antioch New England Graduate School
40 Avon Street
Keene, NH 03431
cthomashow@antiochne.edu

Peter Verbeek
Miyazaki International College
1405 Kano
Kiyotake-cho
Miyazaki 889-1605
Japan
pverbeek@miyazaki-mic.ac.jp

Name Index

Abrams, M. H., 203
Acredolo, L. P., 55
Agnetta, B., 8
Agras, S., 42
Allen, K. M., 188
Alloway, N., 56
Altman, I., 122
Anderson, E., 231, 243
Anderson, J. W., 38, 48
Andrews, G., 36
Andrews, K., 8
Angelino, H., 42
Applebome, P., 285
Applegate, J., 143
Appleton, J., 44, 45
Archer, J., 36
Aries, P., 300
Armstrong, B. T., 45, 52
Atran, S., 4, 68, 71
Attanucci, J., 38
Au, T. K., 84
Augustini-Ziskind, M. L., 75
Aureli, F., 6, 9

Backscheider, A. B., 69
Bacon, L., 167
Bailey, C., 163
Balling, J., 49
Balling, J. D., 229, 230, 252
Bandura, A., 213
Barker, R., 118
Barkow, J. H., 1
Barney, G., 141
Bauer, D. H., 43
Baun, M. M., 188
Beck, A., 188
Beck, U., 290
Behrend, D. A., 68
Berg, M., 125
Bergin, B., 187
Bergstrom, N., 188
Berk, L., 38
Bernhard, J. G., 7, 9, 10
Bernstein, A. C., 73
Berry, T., 299, 300
Berry, W., 290
Bettelheim, B., 135
Bierhorst, J., 71
Billmann-Maheca, E., 163
Birjulin, A. A., 167
Birney, B., 144
Bixler, R. D., 240
Blascovich, J., 188
Bloch, M., 85
Bloom, B. S., 121, 125
Blumberg, F. C., 54
Boesch, C., 8, 14, 17
Bogin, B., 32
Bornstein, M. H., 252
Bower, T. G. R., 37
Bowlby, J., 35, 37
Brannon, C., 53
Briffett, C., 217
Bronfenbrenner, U., 118
Bronson, G. W., 39
Brooks, J., 40, 41

Bruns, G. L., 205
Brunson, L., 219
Bryant, B., 108
Buchanan, C. M., 247
Bullard, R. D., 108
Bunting, T. E., 216
Burgess, J., 217
Burke, E., 294
Byrne, R., 7, 8

Call, J., 5, 8, 12
Carey, S., 67, 68, 69, 70–71, 74, 85, 157
Carlisle, C. L., 240
Carson, R., 127–128, 140–141, 143, 299
Cartmill, M., 192
Cassimatis, N., 76
Charnov, E. L., 9
Chase, R. A., 38
Chawla, L., 125, 203, 204, 207, 212, 213, 214, 215, 216, 217, 219
Checkoway, B., 249
Chipeniuk, R., 56
Chomsky, N., 39
Chu, J., 313
Clarke-Stewart, K. A., 38
Claxton, G., 55
Cobb, E., 127, 214
Cocking, R. R., 53
Cohen, U., 45, 52
Colborn, T., 280
Coleridge, S. T., 203
Coles, R., 10, 138
Coley, J. D., 68, 70, 71, 72, 78, 81, 84, 85
Coley, R. L., 219
Collaris, R., 35, 42, 51
Collins, W. A., 252
Commager, H. S., 293
Comstock, A. B., 309
Cook, E. W., 37, 51
Cornell, J., 126
Corno, L., 54
Corsen, E. O. L., 187
Corsen, S. A., 187
Cosmides, L., 1, 35
Costanza, R., 295
Cousins, L. R., 216
Cowan, P. A., 73
Crider, D. M., 241
Critser, G., 281
Croake, J. W., 42
Crook, J. H., 9
Crooks, C. L., 50, 230

Daly, H., 295
Daly, M., 40
Damasio, A. R., 9
Damon, W., 94
Darwin, C. R., 16
Davey, G. C., 35, 51
Davies, S., 51
De Long, A. J., 45
de Waal, F. B. M., 2, 6, 10, 11, 13, 15, 17, 19
DeCarie, T. G., 40
Derr, V., 126, 136, 137–138, 138, 146
Descartes, R., 9
DeVore, I., 32, 246
DeVries, R., 94
DiPaoli, M., 179
Dolgin, K. G., 68
Dollins, J., 42
Douglas, W. O., 316
Draper, P., 48, 55
Driver, B., 138
Dubow, J., 141
Duda, M., 143
Dumanoski, D., 280
Dunlap, J., 144

Eagles, P. J., 131
Eaves, L. J., 36
Ebata, A. T., 244, 245
Eccles, J. S., 247
Egeland, B., 9
Ehrlich, A., 313
Ehrlich, P., 313
Eiseley, L., 134
Engel, S., 135, 218

Engelhart, M. B., 121
Erikson, Erik, 133
Erikson, M. F., 9
Ewart, A., 138

Faber, D., 108
Fagen, R., 7, 10
Falk, J., 49
Falk, J. H., 229, 230, 252
Fallon, A. E., 75
Feuerstein, G., 207
Field, D. J., 6
Fine, A., 179
Fiorito, E., 110
Fishman, C., 282
Flanagan, C., 247
Flannery, K. V., 251
Floyd, M. F., 240
Fogel, A., 163
Fontaine, A. M., 45
Fouts, R., 16
Frensch, P. A., 54
Friedman, B., 95, 106, 163
Friedmann, E., 179, 188
Funk, R. B., 241
Furst, E. J., 121

Gadamer, H. G., 205–206
Gallimore, R., 184
Garber, P. A., 12, 13
Gardner, W., 9
Garfat, T., 159
Garrett, E., 170
Garvey, C., 38, 40–41
Gates, J., 292, 295
Gebhard, U., 163
Gebser, J., 206–215, 217, 220–221
Gelman, R., 4
Gelman, S. A., 65, 69, 73, 74, 77, 78, 79, 80, 81, 82
Gendlin, E. T., 159
Gibson, E. J., 5, 6, 20
Gibson, J. J., 5
Gieseke, T. J., 167
Gilbert, P., 56
Gilligan, C., 159
Goldschmidt, W., 247–248
Goleman, D., 285
Goodall, J., 2, 13, 14, 17
Goodman, J. F., 187
Gopnik, A., 65
Gould, S. J., 207
Green, S. B., 54
Greenberg, D., 40
Greenfield, P. M., 53, 54
Greenspan, S., 289
Grice, D., 40
Griffiths, M., 53, 54
Groombridge, B., 141
Guba, E. G., 204
Gutheil, G., 77, 82, 84
Guttmann, G., 188

Hagman, E. R., 42, 43
Hahn, U., 82
Hales, S. S., 240
Hamilton, D. I., 36, 41, 42, 44
Hammitt, W. E., 240
Hammond, P. C., 313
Hand, G., 108
Handford, A. G., 54
Hannon, B., 12
Haraway, D. J., 1
Hare, B., 8
Harlow, H. F., 9
Harlow, M. K., 9
Harman, A., 283
Harris, J. R., 243
Harrison, C., 217
Hart, L. A., 187
Hart, R., 43, 45, 46, 48, 52, 55, 215, 216, 217, 218, 252
Hart, R. A., 125, 146
Harvey, M. R., 216
Harvey, P. H., 32
Hatano, G., 68, 69, 70, 73, 77, 157
Healy, J., 282, 283, 285
Heath, A. C., 36
Hebb, D. O., 247
Heerwagen, J. H., 47, 55
Heidegger, M., 205
Heisel, W. J., 35, 42, 44

Heit, E., 82
Hendy, H. H., 187
Herbert, E. J., 50, 230
Hertsgaard, M., 107
Herzog, T. R., 50, 230
Hester, N., 240
Hester, R. T., 218, 240
Hetherington, E. M., 252
Heywood, V., 141
Hickling, A. K., 69
Hill, K., 32
Hill, W. H., 121
Hillman, J. D., 40
Hirschfeld, L. A., 73, 74
Hodes, R. L., 37, 51
Hoelscher, K., 159
Hoffman, E., 215
Holmes, F. B., 37, 42, 43
Holmes, S. J., 201, 221
Holzberg, C. S., 318
Hoogstra, L., 158
Horner, T. M., 40
Howe, D., 95, 163
Hoyt, K. A., 55
Hrdy, S., 40
Huber, P., 107
Huffman, M. A., 15
Hugdahl, K., 37, 51
Humphrey, N. K., 173
Hungerford, H. R., 212
Hurphy, G. C., 42
Hurtado, A. M., 32
Husserl, E., 205
Huston, A. C., 54

Iltis, H., 298
Inagaki, K., 68, 69, 70, 73, 77, 157
Iozzi, L. A., 126

Jaakkola, R. O., 157
James, A., 205
James, L., 72, 85
Jefferson, T., 293
Jenks, C., 205
Jersild, A. T., 37, 43
Jersild, C. L., 43
Johnson, J. H., 237
Johnson, S. C., 73, 74, 79
Joy, B., 297
Jung, C., 120

Kaban, B., 38
Kagan, J., 164
Kahn, P. H., Jr., 3, 5, 95, 97, 105, 106, 111, 122, 125, 163, 170, 173, 204, 211, 309
Kakamura, C. Y., 42, 44
Kalish, C. W., 73, 74, 75, 77, 78
Kals, E., 217
Kano, T., 12
Kaplan, R., 50, 110, 125, 228, 230, 232, 234–235, 238, 246, 251
Kaplan, S., 110, 125, 138, 228, 238, 246, 251
Karker, A. C., 37, 51
Katcher, A., 179, 180, 188
Kaye, D. M., 179
Keil, F. C., 73, 77, 79, 84
Kellert, S. R., 1, 3, 110, 117, 118, 122, 123, 125, 126, 129, 130, 131, 132, 135, 136, 137–138, 142, 143, 144, 163, 219, 315, 316
Kelsey, R. M., 188
Kendler, K. S., 36
Kessler, R. C., 36
Kett, J. F., 243–244
King, N. J., 36, 42, 44
Kippax, S., 54
Kirkby, M., 45, 46, 52, 53, 219
Kitman, J., 281
Knudtson, P., 71
Kocher, S., 56
Kohlberg, L., 94, 95, 111
Kohnstamm, M., 179
Kong, L., 217
Konner, M., 17, 48
Korpela, K., 238
Krathwohl, D. R., 121, 125
Kummer, H., 7
Kuo, F. E., 216, 219
Kuroda, S., 18

Ladd, F., 242–243
Lamb, M. E., 9
Lancaster, M., 240
Lang, P. J., 37, 51
Langley, J. D., 51
Langston, N. F., 188
Lanting, F., 2, 11, 19
Lapouse, R., 42, 44
Larson, R. W., 241
Lasn, K., 297–298
Laughlin, W. S., 250
Laurendeau, M., 68
Lawrence, E., 123, 153
Lazarus, R. S., 9
Leakey, R., 284
Lease, G., 5, 205
Ledewitz, B., 292
Lee, R. B., 32, 246
Legendre, A., 45
Leopold, A., 316
Levi, D., 56
Levi-Strauss, C., 120, 192
Levin, D. T., 77, 84
Levinson, B., 187
Lewin, R., 10, 11, 284
Lewis, C. A., 239
Lewis, M., 40, 41
Liacas, T., 297–298
Lickona, T., 131
Limb, M., 217, 220
Lipsitz, J., 245
Little, B. R., 217, 241–242, 249
Lockard, D., 296
Locke, J., 296
Lockwood, R., 188
Loftus, E., 53
Loftus, G., 53
Lohr, D., 53
Lopez, A., 82
López, A., 81–82
Lorenz, K., 2
Losito, B. D., 110
Losos, E., 141
Lott, D. F., 9
Lourenço, O., 95
Lovelock, J., 4
Lowry, P., 45
Lumsden, D., 129
Lynch, J., 188
Lynch, K., 216, 219, 237–238
Lyons, E., 49

Mabey, R., 319
Maccoby, E. E., 252
MacIver, D., 247
Mader, B., 187
Maestripieri, D., 9
Magen, Z., 247, 249
Mahapatra, M., 164
Maker, C. J., 121, 131
Malone, T. W., 53
Mander, J., 143
Mandinach, E. B., 54
Manley, B., 179
Margadant-van Arcken, M., 157, 163
Markey, A. T., 43
Markey, F. U., 43
Markman, E., 4
Markman, E. M., 81
Marks, I. M., 40, 44, 246
Marquis, J., 54
Maruhashi, T., 18
Masia, B. B., 125
Matthiessen, P., 259
Matson, J. L., 35, 42, 44
Matsuzawa, T., 5, 15
Matthews, H., 220
Matthews, M. H., 45, 48
Matthews, R., 293
Mayer, R., 111
McCormick, M., 77, 78
McGinty, T., 45, 52
McGrew, W. C., 8, 14, 17
McLean, P. D., 9
McNally, M. J., 240
McNally, R. J., 37
Mech, E. V., 42
Medich, E. A., 125
Medin, D., 68, 71, 72, 78, 79, 85
Medina, A. Q., 231, 233
Meloy, E., 109

Melson, G. F., 163
Menzel, C. R., 12
Menzies, R. G., 51
Merckelbach, H., 35, 42, 51
Messent, P., 187, 188
Meyers, J. P., 280
M'Gonigle, M., 286
Mickunas, A., 207
Midgley, C., 247
Midgley, M., 245
Miles, M. A., 110
Miller, J. G., 164
Miller, P. J., 158
Miller, R. A., 46
Milton, K., 13
Mohai, P., 108
Monk, M. A., 42, 44
Montada, L., 217
Montagu, A., 312, 317
Moore, E. O., 110
Moore, G. T., 45, 52
Moore, R., 219
Moore, R. C., 44, 48, 52, 55, 125, 138, 146, 216, 218
Moran, C., 36
Morgan, G. A., 40
Muffitt, S., 131
Mullahey, R., 249
Muris, P., 35, 42, 51
Murray, J. P., 54
Mwanza, N., 18
Myers, G., 100
Myers, N., 141
Myers, O. E., Jr., 155, 156, 158, 159, 160, 163, 170, 171

Nabhan, G. P., 10, 120, 125, 138, 143, 300, 309
Nash, R., 316
Neale, M. C., 36
Neel, J. V., 48
Nevers, P., 163
New, T., 313
Newson, E., 56
Nishida, T., 8, 14, 17
Nishihara, T., 18
Nixon, W., 309
Noddings, N., 162, 163, 164
Norton, B. G., 166

Oates, J. F., 6
Obasanjo, O. O., 241, 243
O'Brien, M., 290
Ogden, T., 192
Öhman, A., 52
Okagaki, L., 54
Oko, R. A., 18
Olds, A. R., 45, 52
Oliveau, D., 42
Ollendick, T. H., 35, 36, 42, 44
Olshausen, B. A., 5–6
Olwig, K. R., 215
O'Reilly, A. W., 81
Orians, G. H., 47, 49
Orr, D. W., 310
Ortony, A., 79
Osherson, D. N., 81–82
Owens, P. E., 237, 240, 242, 249

Packer, M., 164
Pagel, M. D., 32
Palmer, J., 212
Panksepp, J., 9
Partridge, E., 165
Peacock, C., 187
Perin, C., 193
Pfeiffer, J. E., 246
Phillips, A., 141
Piaget, J., 67, 94, 131
Pinard, A., 68
Pinker, S., 35, 39
Poresky, R. H., 163
Porteous, J. D., 217, 218
Postman, N., 291, 297
Poulton, R., 51
Predovic, M., 188
Press, E., 283
Price, J., 248
Prout, A., 205
Putnam, R., 289

Pyle, R. M., 120, 125, 138, 141, 143, 144, 146, 308, 312, 313, 314, 319, 323

Rahn, S., 218
Ratanapojnard, S., 122, 126, 146
Redefer, L. A., 187
Reed, E. S., 5, 7, 14, 20
Regan, T., 9
Reisner, M., 307
Reuman, D., 247
Reynolds, F., 13
Reynolds, V., 8, 13, 14, 17
Ricciuti, H. N., 40
Richards, D. D., 68
Richman, B. A., 77, 84
Ricoeur, P., 204
Rochberg-Halton, E., 159
Romo, L. F., 84
Rosengren, K. S., 77, 78
Ross, M., 179
Ross, N., 68, 71
Ross, S., 179
Rothstein, D., 141
Röver, M., 163
Rozin, P., 75
Russell, P. A., 36, 42, 43
Ryan, R. L., 228, 246

Salapatek, P., 42, 43
Salvadori, I., 219
Sapolsky, R., 10
Saunders, C. D., 3, 4, 167, 170
Savage, J., 141
Savage-Rumbaugh, S., 10, 11
Scarr, S., 42, 43
Schaller, G. B., 248
Scherer, M. W., 42, 44
Schino, G., 9
Schneider, B., 244
Schulenberg, J., 244, 245
Schumacher, D., 217
Scott, J., 286
Searles, H. F., 125, 128, 133, 139, 201
Sebba, R., 128, 139–140
Selden, G., 320
Shafir, E., 81–82
Shakespeare, W., 218
Shanker, S. G., 10
Shapiro, B., 38
Shatz, M., 69
Sheldon, L., 53, 54
Shepard, P., 7, 8, 120, 122, 133, 135, 260, 262, 263, 265, 266, 288, 298–299, 299, 300
Sherwood, K., 220
Shrader-Frechette, K., 296
Shuman, M., 298
Shweder, R. A., 164
Siegler, R. S., 68
Silva, P. A., 51
Simons, R. F., 110
Singer, P., 9
Sinha, C., 20
Sinsheimer, R., 297
Skarin, K., 40
Sloboda, J., 40
Smith, E. E., 81–82, 82
Smith, P. K., 37, 39, 40, 42
Smucny, D., 9
Snyder, S., 312, 317
Soares, J. J. F., 52
Sobel, D., 125, 126, 133, 134, 138, 146, 216, 218, 264, 309
Sodhi, N., 217
Solomon, G. E. A., 73, 74, 76, 79, 85
Solomon, R., 40
Soulé, M. E., 5, 205
Spada, E. C., 6
Springer, K., 73, 74
Sroufe, L. A., 9
St. Antoine, S., 143
Stafford, K., 309
Stavy, R., 68
Stegner, W., 134
Steinberg, L., 252
Stern, D., 156, 157, 159
Stevens, A., 248

Stevenson, D., 244
Stone, C., 316
Stutz, E., 55
Suggate, J., 212
Sugiyama, Y., 8, 14, 17
Sullivan, W. C., 216, 219
Susskind, Y., 249
Suzuki, S., 18, 71
Sylvester, D., 42

Talbot, J., 138
Tapsell, S. M., 216
Taylor, A. F., 216, 219
Taylor, M., 220
Taylor, T. J., 10
Tharp, R. G., 184
Theodorou, P., 164
Thoma, L., 188
Thomas, C. D., 313
Thomas, D., 129
Thomas, J. A., 313
Thomas, W., 187, 188
Thomashow, M., 125, 264, 266
Thompson, R. A., 9
Thompson, W. R., 247
Thornton, J., 280
Titman, W., 216
Tomaka, J., 188
Tomasello, M., 5, 8, 12
Tooby, J., 1, 35
Trimble, S., 10, 120, 125, 138, 300, 309
Troisi, A., 9
Tuan, Y. F., 215
Tucker, F., 220
Turiel, E., 94, 95
Turner, V., 179, 189–190
Tutin, C. E. G., 8, 14, 17

Ulrich, R. S., 11, 55, 56, 110
Untermeyer, L., 127
Urton, G., 153
Uzgiris, I. C., 48

Van Hooff, J., 9
Van Lawick-Goodall, J., 2
Vanderbeck, R. M., 237
Verbeek, P., 6
Vigdor, M. G., 179
Volk, T. L., 212
Vollbracht, C., 144
Voltaire, 9
Vygotsky, L., 184

Wals, A. E. J., 111, 211
Waser, P. M., 19
Washburn, J., 283
Wax, N., 68
Weaver, A. C., 10
Weil, S., 138
Weisfeld, G. E., 9
Weissman, M. D., 74
Wellman, H. M., 65, 73, 80
Wendt, J. S., 251
Westervelt, M. O., 131, 132, 163
White, B. L., 38, 55
Whiten, A., 8, 9, 14, 17
Whitman, W., 127
Wigfield, A., 247
Wiggins, G., 159
Wilcove, D. S., 141
Wiley, A., 219
Wilkie, O., 81
Wilkins, G., 179, 180
Williams, T. M., 54
Willits, F. K., 241
Wilson, E. O., 1, 3, 55, 104, 110, 117, 123, 129, 141, 142, 299, 315
Wilson, M., 40
Winnicott, D. W., 179, 193
Wohlwill, J. F., 4, 5, 122
Wong, H., 219
Wood, L., 3, 4
Wordsworth, W., 199–204
Wrangham, R. W., 8, 14, 15, 17
Wright, J. C., 54

Yabrove, G., 164
Yamagiwa, J., 18
Young, D., 44, 48, 52, 55, 125
Young, K., 143

Yuen, B., 217
Yumoto, T., 18

Zaitchik, D., 74
Zan, B., 94
Zegans, L. S., 39
Zegans, S., 39
Zelson, M., 110
Zemanek, M., 188
Zimmerman, L. K., 211

Subject Index

Adaptive learning rules, 1
Adolescents. *See also* Children
  Australian landscape and, 230–231
  autonomy and, 247–248
  biome preference and, 229–230
  changing trends in, 243–245
  culture and, 245–246
  direct experience with nature, 240
  favorite places, 237–238
  informational needs of, 246–250
  leisure patterns, 240–242
  magic and archaic consciousness and, 217
  management of public land and, 267–270
  photo-based studies on environment preferences and, 230–242
  protection of wildlife sanctuaries and, 270–274
  river corridors and, 232–233, 234–235*f*, 236–237
  rural forests and, 231
  service organizations and, 245
  sexuality and, 262
  snakes and, 259–260
  urban environments and, 231–232, 239–240
  value of nature and, 135–136, 250–252, 263–266
  wilderness experiences and, 136–138, 238–239
  after World War II, 244–245
  zoos and, 259–260, 274–277
Aesthetics, 2, 96–97, 103–104, 107, 108, 127–128, 130, 132–133, 140–141, 142, 202, 215, 240
Affective development, 8–10, 120, 125–129, 287–288, 316
Affordances, 6, 7, 14
Aggressive behavior, 184
Animals
  animate relatedness with children and, 156–157
  animism, 67–69
  anthropocentrism and, 69–72
  avoiding dangerous, 41–43
  children's caring relationships with, 160–168
  children's early social interactions with, 154–156
  children's idea of self and, 159–160
  children's relationship to domestic, 95
  dialogue and sociability of children and, 187–188
  domains of social relatedness with children and, 156–159
  environmental care and, 165–167
  intersubjective relatedness with children and, 157–158
  liminality and, 188–194
  linguistic relatedness with children and, 158–159
  morality toward, 163–165
  natural care of, 160–163

Animals (cont.)
in therapeutic education, 179–187
zoos and, 3, 144–145, 167–168
Animate relatedness, 156–157
affectivity, 156, 160
agency, 156, 160, 212
coherence, 156, 160
continuity, 156, 160
Animism, 67–69, 202
Anthropocentric reasoning, 69–72, 96–99, 103, 111, 166
Archaic consciousness, 208–209, 212–213, 215–217
Attachment theory, 37–39
Attention-deficit hyperactivity disorder (ADHD), 179–180
Australian landscape study, 230–231
Autonomy, 247–248

Biocentric reasoning, 96–99, 166
Biological causality
growth and natural change and, 77–78
illness and, 75–77
inheritance and, 73–75
Biological environments, 30–31
Biome preference, 229–230
Biophilia, 1–5, 9–10, 15, 16, 20–21, 110, 117, 123–124, 129, 173, 179, 299, 315–316
Boys. *See also* Children
care toward nature and, 169–171
caring relationships with animals and, 161
constructing shelter and, 46
protection of wildlife sanctuaries and, 271–272
video games and, 53
Brookfield Zoo, 3, 160, 259
Butterflies, 313–315

Caring in relationships, and natural care, 160–163
Categorization of plants and animals, 30–31, 65
essentialism and, 79–80
inductive potential and, 80–82
Children. *See also* Adolescents; Boys; Girls; Infants
aggressive behavior and, 184
animate relatedness and, 156–157
animism and, 67–69, 202
anthropocentric reasoning, 69–72, 96–99, 103, 111
archaic consciousness and, 212–213
attachment to caregivers and, 9–10
avoidance of animals by, 41–43
avoidance of danger by, 35–37, 52–53
avoidance of strangers by, 39–41
behavior toward peers and, 185–186
behavioral responses to environmental stimuli, 33–50
biocentric reasoning and, 97–99
biological causality and, 72–78
biological inheritance and, 73–75
biophilia and, 1, 3–4, 20–21, 117, 123–124, 129, 173, 179, 299, 315–316
care of nature and, 168–172
caring relationships with animals and, 160–168
contact with nature and, 118–120, 127–129, 138–146
crying and, 36
domains of social relatedness with animals and, 156–159
early social interactions with animals and, 154–156
environmental care for animals and, 165–167
environmental generational amnesia and, 105–112
essentialism and, 79–80
fathers and, 37–39
fear and, 36–37, 41–43, 51–52
finding and constructing shelter and, 44–47, 48–50
folkbiology and, 82–85
food recognition and, 47–48

growth and natural change and, 77–78
health of American, 280–282
the idea of self and, 159–160, 204–206
illness and, 75–77
inductive potential and, 80–82
intersubjective relatedness and, 157–158
learning and, 179–187
liminality and, 188–194
linguistic relatedness and, 158–159
magic consciousness and, 214–217
modern political economy and, 286–292
moral values and knowledge about the environment and, 101–105
morality toward animals and, 163–165
mother-infant attachment and, 37–39
natural habitats of, 323–325
natural play and, 55–56
nature, romantic connections between and, 199–204
nature-deprived, 315–317
obesity in, 281–282
outdoor play and, 45–47, 52–53, 145–146, 215–217, 311–312
pastoralism and, 216
plant and animal categorization and, 78–82
pollution and, 95–97, 106–107
reasoning skills of, 70–71, 84–85
relationships to domestic animals, 95
rights of, 292–300
saving special places and, 320–325
shared responsibility for raising, 32–33
sociability of, 187–188
sympathy and, 204–206
television viewing and, 54–55, 282–283, 317–318
therapeutic education and, 179–187
topographic hazards and, 43–44
value of nature and, 132–133, 305–325
video games and, 53–55, 283–284
violence committed by, 279–280
William Wordsworth on, 199–204
zoos and, 180–182
Climate and ecology, 284–286
Cognitive development, 120–125, 133–134, 289–290, 315
Communitas, 189–190
Computers, 56–57, 119–120, 140, 143–144, 282–284, 299, 315, 317–319, 325
Consciousness
archaic, 208–209, 212–213, 215–217
magic, 209–210, 214–217
mythic and mental experience and, 210–212, 217–220
Conservation
of land/species, 109, 168, 292, 321
support for, 136–137, 312, 314–315
Creative destruction, 287–288
Culture
animism and, 68
anthropocentrism and, 71–72
evolution and, 245–246
folkbiology and, 84–85
moral values and knowledge and, 101–105
preferred environments and, 228–229

Danger avoidance, 10–11, 35–37
Deconstructionism, 5, 204–206
Descent of Man, The (Darwin), 16
Development
affective, 8–10, 120, 125–129, 287–288, 316
affiliations with nature and, 94–101, 168–172, 212–213, 250–252
animism and, 67–69, 260–266
behavioral responses to environmental stimuli and, 33–50

Development (cont.)
biophilia and, 1, 3–4, 20–21, 117, 123–124, 129, 173, 179, 299, 315–316
cognitive, 120–125, 133–134, 289–290, 315
ecological-evolutionary perspective, 29–33
evaluative, 120–121, 129–138, 163–167, 299–300, 316
folkbiology and, 65–67
liminality and, 189–194
physical, 6–7, 32–33, 315
Direct experience with nature, 118–119, 123–124, 138–146, 240
Disequilibration, 94, 99

Ecological identity, 263–267
Ecological Psychology (Barker), 118
Ecology
adolescent ecological thinking and, 260–263
children's behavioral responses to environmental stimuli and, 33–50
climate and, 284–286
foraging and feeding and, 47–48
modern life and, 50–56, 119–120
physical and biological environments and, 30–31
rights of children and, 292–300
Ecology of Human Development, The (Bronfenbrenner), 118
Ecology of Imagination in Childhood, The (Cobb), 127
Economy, political, 286–292
Education
commercialization of, 283
ecological identity and, 263–267
natural history and, 309–311
structural-developmental (constructivist) approach to, 110–112
Emotions
of humans, 8–10
mythic places and, 217–220
nature and, 1, 127–129
of primates, 8–10, 13
Environmental generational amnesia, 105–112. *See also* Extinction of experience
Environments. *See also* Nature
adolescents' preference for natural, 250–252
anthropocentric reasoning about, 69–72, 96–99, 103, 111, 166
archaic identity and, 212–213
biocentric reasoning about, 97–99, 166
biological, 30–31
biome preference of adolescents and, 229–230
children's behavioral responses to, 33–50
classification of, 30–31, 65, 78–82
cross-cultural comparisons of values and knowledge about, 101–105
fear and, 33–44, 51–52
finding and constructing shelter and, 44–47
generational amnesia about, 105–112
information and, 29–30, 33–34, 35, 112, 173, 228, 250–251
physical, 30
pollution and, 95–97, 103, 106–107
preferred, 228–229
river corridor, 232–233, 234–235*f*
rural forest, 231
structure and development of children's conceptions about, 94–101
urban, 231–232, 239–240, 284–286, 305–306, 311–312
Essentialism, 79–80
Evaluative development, 120–121, 129–138, 163–167, 299–300, 316
Ever-Present Origin, The (Gebser), 206–208
Evolution
biological environments and, 30–31
consciousness and, 207–208

culture and, 245–246
ecological environments and, 29–30
hunter-gatherer social groups and, 31–32, 35
modern life and, 50–56
physical environments and, 30
physiological and social environments and, 31–33
Extinction of experience, 141, 312–315. *See also* Environmental generational amnesia

Facial expressions, 3
Facilitation, 18–19
Fathers, 40
Fear
of animals, 41–43
of being alone, 43
of darkness, 42–43
modern life and, 51–52
of possible danger, 36–37
of strangers, 39–41
of the supernatural, 42–43
topographic hazards and, 43–44
Folk theories, 65
Folkbiology
animism and, 67–69
culture and, 84–85
definition of, 65–67
development of, 82–85
early, 67–82
growth and natural change and, 77–78
illness and, 75–77
reasoning and, 84–85
Freedom and ecological rights, 294–295

Gender differences, 39–41, 46, 53, 161–162, 169–171, 245, 271–272, 316
Girls. *See also* Children
care toward nature and, 169–171
caring relationships with animals and, 161–162
constructing shelter and, 46
video games and, 53

Handbook of Nature Study (Comstock), 309
Hermeneutics, 204–206
Humans. *See also* Children
childhood memories of nature and, 305–310
emotional nature of, 9–10
hunter-gatherer social groups, 31–32, 35
liminality and, 188–194
mother-infant attachment and, 37–39
mythic places and, 217–220
recognition and categorization of living things, 3–4
response to primates, 2–5
response to various landscapes, 48–50
visual perception, 5–6
Hunter-gatherer social organization, 31–32, 35

Illness
folkbiology and, 75–77
threats to health, 280–282
Indirect experience with nature, 119
Inductive potential, 80–82
Infants. *See also* Children
chemicals and, 280–282
defense mechanisms of, 37
fear of strangers, 39–41
human versus ape, 32–33
mother attachment of, 37–39
mouthing small objects and, 47–48, 51
vision, 38
Information
affective development and, 125, 126
cognitive development and, 122, 123
environmental, 29–30, 33–34, 35, 112, 173, 228, 250–251
folkbiology and, 65, 83

Information (cont.)
human needs and, 246, 249–251
television as source of, 54, 282–283
Inheritance, biological, 73–75
Intellectual development. *See* Cognitive development
Intersubjective relatedness, 157–158
Intimacy, nature, 312
Isomorphism, 98–99, 102
Israel, 68

Japan, 68
Japanese monkeys, 12

Kahuzi-Biega National Park, 19
Knowledge
development and, 93–94, 103, 110–111, 113–114, 121–123, 127–128, 132–133
environmental, 5, 9, 12, 48, 56, 65, 69, 72, 101,113, 157, 166–167, 168, 170, 228, 250–251, 272, 287, 310, 311, 312
folkbiology and, 66, 70, 71–72, 74, 76, 77, 81, 82–83, 84, 85–86

Language, and children's relatedness with animals, 158–159, 174*n*1
Lead poisoning, 281
Learning
animals in therapeutic education and, 179–187
by emulation, 8, 15
by imitation, 8, 10, 15
liminality and, 190–193
social, 7–8
stimulus enhancement, 7–8
Liminality, 188–194
Linguistic relatedness, 158–159
Literacy, nature, 312
Lyrical Ballads (Wordsworth), 204

Magic consciousness, 209–210, 214–217
Man and the Computer (Montagu and Snyder), 312
Materialism, 287–289
Medicinal plant use by primates, 15–16
Menominee Indians, 71–72
Morality
toward animals, 163–170, 172–174, 181–183, 186–187, 191–193
biophilia and, 173–174
children's values of nature and, 93, 94–105, 132, 135–136
nature and, 316–317
values-related development and, 129–131
Mother-infant interactions
in humans, 37–39
in primates, 3
Mythic and mental experience, 210–212, 217–220

National Outdoor Leadership School (NOLS), 136
Native Americans, 68, 71–72
Nature. *See also* Environments
adolescents' preference for, 135–136, 250–252, 263–266
affordances of, 6, 7
biological diversity and, 142
children deprived of, 315–317
children's care toward, 168–172
computers and, 56–57, 119–120, 140, 143–144, 282–284, 299, 315, 317–319, 325
direct experience with, 118–119, 123–124, 127–129, 138–146, 240, 305–309
emotions and, 9–10, 127–129
environmental generational amnesia and, 105–112
exploitation of, 7, 8, 11–12
exploration of, 6–7, 11–12
extinction of experience and, 312–315
hermenutics and, 204–206
indirect experience with, 119
intimacy, 312
kinship with, 16–19

knowledge formation about, 121, 122–123
literacy, 312
management of public land and, 267–270
manipulating, 14–16
materialism and, 286–289
pastoralism of children and, 216
saving special places and, 320–325
sense of wonder and, 16–18
statistical similarity in, 5–6
symbolic experience with, 119–120
vicarious experience with, 119–120, 123, 143–144

Obesity in children, 281–282
Our Common Future (World Commission on Environment and Development), 294
Outdoor play, 45–47, 52–53, 215–217, 311–312
Outward Bound (OB), 136

Pastoralism, 216
Perception, visual, 5–6
Physical environments, 30–31
Play
magic and archaic consciousness and, 215–217
natural, 55–56
outdoor, 45–47, 52–53
vacant lots and, 311–312
video games and, 53–55
Politics
children's exposure to nature and, 286–292
management of public land and, 267–270
protection of wildlife sanctuaries and, 270–274
rights of children and, 292–300
saving special places and, 320–325
Pollution, 95–97, 103, 106–107
Prelude, The (Wordsworth), 204
Primates
avoidance of danger and, 10–11, 40
deriving pleasure from nature and, 16
emotional nature of, 8–10, 13
exploration and exploitation and, 6–7, 8, 11–12
interactions between, 4–5, 18–19
medicinal plant use by, 15–16
physical development, 32–33
sense of wonder about nature and, 16–18
social nature of, 7–8
tool use by, 14–15
Public land, management of, 267–270

Rachel Marshall Outdoor Learning Laboratory, 270
Reasoning
anthropocentric, 69–72, 96–99, 103, 111, 166
biocentric, 96–99, 166
about biological phenomena, 70–71
categories of conceptions of harmony and, 100–101
environmental, 104–105
folkbiology and, 84–85
isomorphism and, 98–99
transmorphism and, 99
Reflections on the Revolution in France (Burke), 294
Relationships
between animals and children, 160–168
children's idea of self and, 159–160
River corridors, 232–233, 234–235*f*, 236–237
Romanticism, 199–204
Rural forests, 231

Safety, children's mechanisms for, 35–37, 52–53
Saving special places, 320–325
Self, idea of, 159–160, 204–206
Sexual behavior
in adolescents, 262

Sexual behavior (cont.)
in primates, 3
Shelter, finding and constructing, 44–47, 48–50
Snakes, 10–11, 259–260
Sociability of children
effect of animals on, 187–188
liminality and, 188–194
Social nature of primates, 7–8
Special places, saving, 320–325
Spots of time, 200
Stimulus enhancement, 7–8
Stranger avoidance by children, 39–41
Structural-developmental theory, 94–95, 110–112
Sympathy, 127–128, 201–202, 205, 210–211, 221

Television viewing, 54–55, 282–283, 317–318
Therapeutic education, animals and children in, 179–187
Transmorphism, 99, 103
Thunder Tree: Lessons from an Urban Wildland, The (Pyle), 308, 319
Tool use by primates, 14–15
Topographic hazards, 43–44
Tucker's Countryside (Selden), 320

Urban environments, 231–232, 239–240, 284–286, 311–312

Vacant lots, 305–306, 311–312
Values, 3, 55, 56, 93–94, 101, 105, 107, 110–111, 112–114, 125–126, 129–136, 166–167, 171, 173–174, 195–196, 217
Vicarious experience with nature, 119–120, 123, 135–136, 143–144
Video games, 53–55, 120, 283–284
Violence
toward children, 40
committed by youths, 279–280
video games and, 53, 56
Vision
adolescents' environmental preferences and, 228–236
infant, 38
perception of nature and, 5–6

Wilderness, 109–110, 114, 136–138, 238–239, 271, 274, 316
Wildlife sanctuaries, 270–274
Wildness, 119, 288, 300, 306, 323–324

Yerkes Regional Primate Center, 17

Zoos, 3, 144–145, 167–168, 180–182, 259–260, 274–277